T0135730

Diss. ETH No. 20043

Mapping Polygons

A dissertation submitted to
ETH ZURICH

for the degree of
DOCTOR OF SCIENCES

presented by
YANN DISSER
Master of Science, TU Darmstadt
Dipl. Inform., TU Darmstadt
born March 4, 1983 in Frankfurt a.M., Germany

accepted on recommendation of
Prof. Dr. Peter Widmayer, ETH Zurich
examiner
Prof. Dr. David Peleg, Weizmann Institute of Science
co-examiner
Dr. Jérémie Chalopin, LIF, CNRS et Aix-Marseille Université
co-examiner
Dr. Matúš Mihaľák, ETH Zurich
co-examiner

2011

Bibliografische Information der Deutschen Nationalbibliothek

Die Deutsche Nationalbibliothek verzeichnet diese Publikation in der Deutschen Nationalbibliografie; detaillierte bibliografische Daten sind im Internet über http://dnb.d-nb.de abrufbar.

ISBN 978-3-8325-3023-5

Logos Verlag Berlin GmbH
Comeniushof, Gubener Str. 47,
10243 Berlin
Tel.: +49 (0)30 42 85 10 90
Fax: +49 (0)30 42 85 10 92
INTERNET: http://www.logos-verlag.de

Abstract

This thesis is concerned with simple agents that move from vertex to vertex along straight lines inside a simple polygon with the goal of reconstructing the visibility graph. The visibility graph has a node for each vertex of the polygon with an edge between two nodes if the corresponding vertices see each other, i.e., if they can be connected by a straight line inside the polygon. While at a vertex, the agent perceives all vertices visible to its current location in the order in which they appear along the boundary. In each step, the agent can choose one of these vertices and move there.

We show that an agent that can distinguish whether two visible vertices are neighbors on the boundary cannot always solve the visibility graph reconstruction problem when restricted to moving along the boundary only. This even remains true if the agent knows the total number of vertices beforehand and if it can measure the angles formed by the boundary of the polygon. On the other hand, we show that an agent that can measure the angles between edges of the visibility graph can always solve the visibility graph reconstruction problem, even when restricted to moving along the boundary only.

We further consider an angle-type sensor which allows to distinguish whether the angle between any two edges is convex or reflex, and a look-back sensor which allows the agent to move back to where it came from in its last move. We show that an agent equipped with both sensors can always solve the visibility graph reconstruction problem, even without prior knowledge about the total number of vertices. The same is true for an agent that can measure the angle between any two edges and has a compass.

For agents that have knowledge of an upper bound on the total number of vertices, we show stronger results. We show that in this setting an agent with look-back sensor or angle-type sensor can always solve the visibility graph reconstruction problem. We further show that multiple, identical, deterministic, indistinguishable such agents can find each other in any polygon.

Zusammenfassung

Diese Arbeit beschäftigt sich mit Agenten, die sich von Eckpunkt zu Eckpunkt entlang gerader Linien innerhalb eines Polygons bewegen, mit dem Ziel den Sichtbarkeitsgraphen zu rekonstruieren. Dabei ist der Sichtbarkeitsgraph der Graph mit einem Knoten für jeden Eckpunkt des Polygons und einer Kante zwischen je zwei Eckpunkten, die sich sehen, deren geradlinige Verbindung also vollständig innerhalb des Polygons liegt. Während sich der Agent an einer Ecke befindet, nimmt er alle von dort sichtbaren Ecken in ihrer Reihenfolge entlang des Polygonrands wahr. In jedem Schritt kann der Agent einen dieser Eckpunkte wählen und sich dort hin begeben. Wir konzentrieren uns auf das Sichtbarkeitsgraphrekonstruktionsproblem, bei dem ein Agent in einem anfangs unbekannten Polygon den Sichtbarkeitsgraphen bestimmen muss.

Wir zeigen, dass Agenten, die unterscheiden können, ob zwei sichtbare Eckpunkte Nachbarn entlang des Rands sind, das Sichtbarkeitsgraphrekonstruktionsproblem nicht immer lösen können, wenn sie sich nur entlang des Polygonrands bewegen dürfen. Dieses Ergebnis bleibt auch dann bestehen, wenn die Agenten zusätzlich die Zahl der Eckpunkte kennen und die Winkel am Polygonrand messen können. Andererseits zeigen wir, dass Agenten, die den Winkel zwischen zwei beliebigen Kanten messen können, das Sichtbarkeitsgraphrekonstruktionsproblem immer lösen können, selbst wenn sie sich nur entlang des Polygonrands bewegen dürfen.

Wir zeigen, dass Agenten, die unterscheiden können, ob der Winkel zwischen zwei beliebigen Kanten konvex oder konkav ist und die sich immer zu ihrer letzten Position zurück begeben können, das Sichtbarkeitsgraphrekonstruktionsproblem immer lösen können, selbst ohne jegliches anfängliche Wissen über die Gesamtzahl der Ecken.

Wir zeigen weiterhin, dass Agenten, die sich immer zu ihrer letzten Position zurück begeben können und denen eine obere Schranke für die Gesamtzahl der Ecken bekannt ist, das Sichtbarkeitsgraphrekonstruktionsproblem immer lösen können. Wir zeigen, dass mehrere, deterministische, identische und ununterscheidbare solche Agenten sich in jedem Polygon gegenseitig finden können. Außerdem zeigen wir, dass beide Resultate auch für Agenten gelten, die unterscheiden können, ob der Winkel zwischen zwei beliebigen Kanten konvex oder konkav ist und denen eine obere Schranke auf die Gesamtzahl an Ecken bekannt ist.

Acknowledgements

I wish to thank everybody who contributed directly or indirectly to this thesis.

First and foremost, I am deeply grateful to my supervisors Peter Widmayer and Matúš Mihaľák for their friendship, support, guidance, patience and expertise.

I thank Jérémie Chalopin for his many contributions to the results presented in this thesis and for acting as a co-examinor.

My thanks also go to David Peleg for agreeing to co-examine this thesis.

A big "thank you" to Gaia Pigino for designing the beautiful cover artwork.

Contents

Chapter 1.

Introduction

Autonomous, mobile robots are taking over more and more tasks that have traditionally been performed by humans. Advances in microelectronics have made robots affordable which are able to perform tasks that we prefer not to do ourselves. A common example are robots that do household chores for us, like cleaning floors or mowing lawns. Some tasks, like demining of land mines, are dangerous and thus best performed by robotic agents. In other situations, autonomous robots are better qualified than humans, especially when a constant awareness level is required, as in guarding or surveillance duties.

Robots for the mass market usually have quite a simple design in terms of hardware. Many cleaning robots, for instance, rely on contact sensors only and make more or less random movement decisions rather than using sophisticated hardware in order to be able to optimize their trajectory. While simple hardware often requires robots to employ somewhat inefficient movement strategies, a simplistic design has many advantages. Foremost, of course, a simple hardware makes robots cheap and thus accessible for the mass market. Cheap robots can be deployed in large numbers for tasks which require robotic coordination, like guarding. In addition, simplistic sensors are generally more robust in terms of measurement inaccuracies and hardware defects. A simple design makes it possible for laymen to deal with maintenance of the robot, which is important for household robots.

The complexity of a robot model can mainly vary in three regards: the sophistication of the robot's movement capabilities, the sophistication of its sensors, and the sophistication of its communication model when multiple robots have to coordinate. The sophistication of the robot's computational power, on the other hand, is usually not a limiting factor due to the availability of powerful and cheap microelectronic components.

Evidently, robots targeted for a certain task cannot be made arbitrarily simple. Some sophistication is needed in order to be able to solve the task. It is a natural question how much complexity is needed for a given task. The aim is to design robots that are weakest possible but still able to solve the task at hand. In general, for a given task, there might be different such minimal designs. An ultimate goal would be to develop a catalogue which lists minimal robot designs required for all common tasks that an autonomous, mobile robot might face. Given one or multiple tasks, this catalogue could then be used to comfortably select a robot design that suffices for solving the given tasks, and that suits other requirements like the cost or availability of the individual components. Ideally, this catalogue would also differentiate how efficiently a task can be solved with different robot designs. This thesis aims to provide first steps towards such a catalogue.

In order to make robotic designs easily comparable, it makes sense to establish a very basic theoretical robot model as well as additional atomic capabilities with which the basic model can be configured. Also, the environment of the robot needs to be captured by a theoretic model. While the resulting model may be not entirely realistic anymore, it allows a more rigorous analysis than a fully realistic counterpart. Still, results for a theoretic model can provide a reference for a realistic design. From now on, we distinguish realistic *robots* from theoretical *agents*.

Before defining an agent model, we have to choose a theoretical representation of the environment in which the robot operates. We could either stay closer to the real-world scenario and model the environment geometrically in two or even three dimensions, or assume a structurally simpler combinatorial approach and assume the environment to be, for instance, a graph. In a geometrical setting, the environment could be bounded by a curve, polygon, etc., or the environment could be unbounded. We could allow obstacles inside the environment, again using our choice of geometric primitives. In a graph-like setting, we might restrict the structure of the graph, for example by assuming the graph to be planar. In this thesis, we go with an environment model that combines both geometrical and graph-like aspects.

We model the environment as a simple polygon without obstacles, but restrict the movements of an agent to be along the edges of the visibility graph of the polygon. More precisely, we assume that an agent moves from vertex to vertex inside the polygon, along straight lines that we call *lines of sight*. While located at a vertex, the agent can observe the polygon locally through its sensors. In order to make informed, local movement decisions, the agent needs a way to distinguish the vertices it sees, i.e., the vertices which can be connected to its current location via a straight line inside the polygon. We provide a means of distinguishing visible vertices by allowing the agent to sense the order in which they appear along the boundary of the polygon. We later equip this basic model with various additional sensors, e.g., for measuring angles between lines of sight. We do generally not impose limitations on the computational power or the amount of memory that an agent possesses.

This thesis mainly focuses on the problem of mapping unknown environments – simple polygons in our case. This problem lies at the heart of many complex tasks and requires an agent to explore an initially unknown polygon with the goal of drawing a map. The map of the polygonal environment for us will always be its visibility graph, i.e., the graph that has a node for every vertex of the polygon and an edge for every line of sight between two vertices. Throughout this thesis, we analyze different extensions of the above basic agent model with the goal of deciding whether the resulting agent can always map its environment or not. For some extensions, we will briefly discuss how multiple, identical agents can coordinate. In scenarios with many agents a fundamental problem for the agents is how to find each other deterministically. Throughout the thesis, we present some results for this so-called *rendezvous* problem.

1.1. Results

We now give a brief and intuitive overview over the results and techniques of this thesis. The exact model will be made formal in the next chapter. For an outline in terms of formal definitions, consider Tables 10.1 and 10.2 in Chapter 10.

The thesis is split into two parts. The first part considers agents that move only along the boundary of the environment. Roughly speaking, the data available to such agents is easy to collect systematically. Therefore, the main challenge of the mapping problem becomes to understand how to use this data algorithmically. The results we develop in this part of the thesis mainly require geometrical reasoning and intuition. In Chapter 4, we consider agents equipped with a combinatorial sensor that allows to distinguish whether two visible vertices are neighbors along the boundary. We show that such agents cannot always draw a map of the polygon, even if they can measure the angles formed by the polygon boundary at their current location. The proofs in this chapter are by giving examples for polygons with different visibility graphs that cannot be distinguished in terms of the data obtained from the agent's sensors. In Chapter 5 we consider agents that can measure the angles between lines of sight. We first show that such agents can always draw a map if they know the number of vertices beforehand. As a proof we give an algorithm that reconstructs the visibility graph from the data collected during a tour around the boundary. We then extend this result to agents that have no prior information about the number of vertices. Essentially, this is because the algorithm from before can be adopted to collect additional data only when it is actually needed.

In the second part of the thesis, we consider agents that can move along any line of sight. We interpret this scenario in the more general context of exploring arc-labeled graphs with certain properties. Chapter 6 considers agents that can distinguish whether the angle formed by two lines of sight is convex or reflex and that can go back the way they came after a sequence of moves. We show that such agents can always draw a map even without any initial knowledge about the total number of vertices. We show that the same is true for agents that can measure angles between lines of sight and have a compass. In Chapter 7 we use techniques from distributed computing in networks of processors in order to develop general methods for the exploration of arc-labeled graphs with an agent that can read arc labels and knows an upper bound on the number of vertices. We show that in this setting agents can always systematically collect all data available to them and we give an algorithm to accomplish this. We establish a certain structure of arc-labeled graphs which allows us to expose useful properties. We show that agents can generally find each other in any graph that admits this structure. In Chapter 8 we direct our attention back at the exploration of visibility graphs. We show that a visibility graph labeled according to the capabilities of agents which can always move back the way they came from admits the desired structure from Chapter 7. Using our general tools, we are able to deduce that such agents can always construct a map, and that multiple such agents can always find each other in the polygon. We then develop an alternative algorithm for collecting the required data, which allows us to improve the algorithm to a polynomial running time overall. Finally, in Chapter 9, we consider agents that can distinguish whether the angle formed by two lines of sight is convex or reflex. We are again able to show that the corresponding visibility graphs admit the desired structure, and use our general tools to show that multiple such agents can always find each other. We then go on to show that the data available to the agent always suffices to construct a map.

It remains an open problem whether the basic agent model without additional sensors already enables agents to always construct a map. In Chapter 10, we give some starting points for further investigations of this question. In particular, we show that convex polygons can always be mapped.

Chapter 2.

Preliminaries

In this chapter we introduce the formal foundation for the thesis. In the process, we summarize well-known properties of polygons and visibility graphs. While we sometimes give alternative proofs for known results, none of the definitions and properties in this chapter are original contributions of this thesis.

2.1. Notational Conventions

Throughout this thesis we use the following notational conventions. We use $[n]$ to denote the set of integers $\{1, 2, \ldots, n\}$ and $\binom{S}{k}$ to denote the set of subsets of size k of a set S. We refer to k-tuples as *sequences* of length k. Let $L = (e_1, e_2, \ldots, e_k)$ and $L' = (e'_1, e'_2, \ldots, e'_{k'})$ be two sequences and let S be a set. We use the following notations. By $|L| := k$ we denote the length of L, by $L_i := e_i$ we denote the i-th element of L, and by $e \in L$ we denote that $L_i = e$ for some $i \in [k]$. By $L \oplus L' := (e_1, e_2, \ldots, e_k, e'_1, e'_2, \ldots, e'_{k'})$ we denote the concatenation of L and L', and we say that L is a *prefix* of $L \oplus L'$. If there exist indices $1 \leq i_1 < i_2 < \ldots < i_k \leq k'$ such that $L_j = L'_{i_j}$ for all $j \in [k]$, we write $L \subseteq L'$ and say that L is a *subsequence* of L'. By $L \cap S$ we denote the longest subsequence of L that contains only elements of S. By $L \backslash L'$ we denote the longest subsequence of L that contains no elements of L'. Let $L^=$ be the longest sequence that is a prefix of both L and L', and let $k^= = |L^=|$ be its length. We say that L is *lexicographically smaller* than L' (with respect to a partial order '$<$' on the elements of L and L') if $k^= < k'$ and either $k = k^=$ or $L_{k^=+1} < L'_{k^=+1}$. We say that L is *periodical* with period p, if $p \leq k/2$, p divides k, and $L_i = L_{i+l \cdot p}$ for all $i \in [k]$ and all $l \in [k/p]$. If L is periodical with period p, we say for every $i \in [k]$ that the elements of $\{L_{i+l \cdot p} | 0 \leq l < k/p\}$ are *periodical partners*.

2.2. Polygons

We will define a polygon in terms of triangles. This approach allows us to quickly derive properties of polygons which will be used throughout the thesis. From now on all geometric considerations are in the plane, and we speak of points meaning elements of $\mathbb{R}^2$.

convex combination

Definition 2.1. The *convex combination* of the points $p_1, p_2, \ldots, p_n$ is the set of points p that can be expressed in the form

$$p = \sum_{i=1}^{n} w_i p_i,$$

where $w_i \in \mathbb{R}, w_i \geq 0$ for each $i \in [n]$, and $\sum_{i=1}^{n} w_i = 1$.

line segment

Definition 2.2. The *line segment* $\overline{pq}$ is the convex combination of two distinct points p and q. We refer to p and q as the *endpoints* of $\overline{pq}$.

triangle

Definition 2.3. A *triangle* $\triangle_{pqr}$ is the convex combination of three distinct points p, q, r that is not a line segment. We call p, q, r the *vertices* and $\overline{pq}, \overline{qr}, \overline{rp}$ the *edges* of $\triangle_{pqr}$.

We now define a polygon in terms of triangles. Note that, for the sake of a unified presentation, we assume no three points of a polygon to be collinear.

polygon

Definition 2.4. We define a *polygon* recursively:

1. Every triangle is a polygon.
2. Let $\mathcal{P}_1, \mathcal{P}_2$ be two polygons and e be an edge of both $\mathcal{P}_1$ and $\mathcal{P}_2$ with $e = \mathcal{P}_1 \cap \mathcal{P}_2$. If no three vertices of $\mathcal{P}_1$ and $\mathcal{P}_2$ lie on one line, then $\mathcal{P} = \mathcal{P}_1 \cup \mathcal{P}_2$ is a polygon. Every vertex of $\mathcal{P}_1$ or $\mathcal{P}_2$ is a vertex of $\mathcal{P}$ and every edge of $\mathcal{P}_1$ or $\mathcal{P}_2$ except for e is an edge of $\mathcal{P}$.

Definition 2.5. The *boundary* of a polygon is the union of its edges, the *interior* of a polygon is the polygon without its boundary.

boundary and interior

We can now show that our definition of a polygon in terms of triangles is equivalent to the standard definition in terms of polygonal curves.

Definition 2.6. Consider a sequence of m points $p_1, p_2, \ldots, p_m$ and the curve $\mathcal{C}$ along the sequence of line segments $\overline{p_1p_2}, \overline{p_2p_3}, \ldots$. If $\mathcal{C}$ does not self-intersect except in p_1 and p_m, we call it a *polygonal curve.* If $p_1 = p_m$ is the only self-intersection of $\mathcal{C}$, then $\mathcal{C}$ forms a *closed polygonal curve.* The points $p_1, p_2, \ldots, p_m$ are the *vertices* of $\mathcal{C}$, p_1 and p_m are the *endpoints* of $\mathcal{C}$, and $p_2, p_3, \ldots, p_{m-1}$ are the *interior vertices* of $\mathcal{C}$.

polygonal curve

We will need the following classical result that formalizes the intuition behind closed curves.

Theorem 2.7 ([33, 52])**.** *A closed curve that does not self-intersect separates the plane into two connected components with the curve as their common boundary. The component containing the point (∞, ∞) is referred to as the* exterior *and the other component as the* interior *of the curve.*

Proposition 2.8. *The boundary of every polygon is a closed polygonal curve, and every closed polygonal curve is the boundary of a polygon.*

Proof. For the first part of the proof we claim that every vertex of a polygon $\mathcal{P}$ is the endpoint of exactly two of its edges. It then follows that the boundary of a polygon is a closed curve. By definition of a polygon, the boundary does not self-intersect, which completes the proof of the first part of the statement. We still need to prove our claim. We do this by induction on the definition of $\mathcal{P}$. If $\mathcal{P}$ is a triangle, then trivially every vertex of $\mathcal{P}$ is an endpoint of exactly two of its edges. Otherwise, let $\mathcal{P}_1, \mathcal{P}_2, e$ as in Definition 2.4 and consider any fixed vertex v. Without loss of generality, assume v is a vertex of $\mathcal{P}_1$. Then, by induction, there are exactly two edges a, b of $\mathcal{P}_1$ which have v as an endpoint. If $e \notin \{a, b\}$, both a and b are edges of $\mathcal{P}$ and the claim holds for

v. Otherwise, v is an endpoint of e and hence must be a vertex of $\mathcal{P}_2$, too. Therefore, by induction, there is a second edge $c \neq e$ of $\mathcal{P}_2$ which contains v. Together, v is the endpoint of the edges $\{a, b, c\}$ in $\mathcal{P}_1$ and $\mathcal{P}_2$ combined. By definition, in $\mathcal{P}$, the set of edges with v as an endpoint is $\{a, b, c\} \setminus e$ and thus has size two. This concludes the proof of the claim as for any choice of v we showed that exactly two edges of $\mathcal{P}$ have v as an endpoint.

For the second part of the proof we consider any closed polygonal curve $\mathcal{C}$ and show that it is the boundary of a polygon. We prove this by induction on the number of line segments in the definition of $\mathcal{C}$. A closed polygonal curve of three line segments forms the boundary of a triangle, which by definition is a polygon. Now assume that $\mathcal{C}$ consists of $k > 3$ line segments. First note that $\mathcal{C}$ is a closed curve and hence, by Theorem 2.7, it separates the plane into interior and exterior. Let v be an endpoint of two line segments s, t of $\mathcal{C}$ that form an angle smaller than π in $\mathcal{C}$. Let u, w be the other endpoints of s and t, respectively. Consider the line segment $\overline{uw}$. We distinguish two cases. First, assume $\overline{uw}$ does not intersect the exterior of $\mathcal{C}$. Then we can define a new closed polygonal curve $\mathcal{C}'$ that uses the line segment $\overline{uw}$ instead of s and t. By induction, $\mathcal{C}'$ is the boundary of a polygon $\mathcal{P}'$. We can identify $\mathcal{P}_1 = \mathcal{P}'$, $\mathcal{P}_2 = \triangle_{uvw}$, $e = \overline{uw}$ in Definition 2.4, and deduce that since $\mathcal{C}$ is the union of the boundaries of $\mathcal{P}_1$ and $\mathcal{P}_2$ without e, it is indeed the boundary of a polygon. Now, assume $\overline{uw}$ intersects the exterior of $\mathcal{C}$. This means that there are points of $\mathcal{C}$ in the interior of $\triangle_{uvw}$. Let z be the point closest to v among them. The line segment $\overline{zv}$ is contained in $\triangle_{uvw}$ and, by definition of z, it does not intersect the exterior of $\mathcal{C}$. We split $\mathcal{C}$ at z and v, and close the two resulting polygonal curves each with the line segment $\overline{zv}$. By induction, both curves are the boundary of a polygon and setting $e = \overline{zv}$ in Definition 2.4 gives us that $\mathcal{C}$ in turn is the boundary of a polygon. □

From now on, we adopt the following conventions whenever we consider a polygon $\mathcal{P}$. By Proposition 2.8, the boundary of $\mathcal{P}$ is a closed curve and we can thus fix an (arbitrary) orientation of the boundary which we will call the *boundary order* of $\mathcal{P}$.[1]

[1] In all illustrations and examples we will use the intuitive "counter-clockwise" order along the boundary as our fixed orientation.

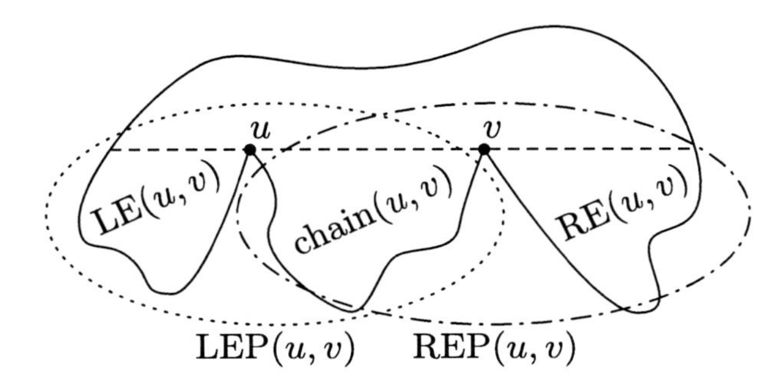

Figure 2.1.: Illustration of RE/LE and REP/LEP. Observe that in this example $\mathrm{RE}(v, u) = \mathrm{LE}(v, u) = \emptyset$.

Also, we fix some (arbitrary) vertex and call it v_0. We denote the number of vertices of $\mathcal{P}$ by n and speak of the *size* of $\mathcal{P}$ when we mean n. We denote the vertices of $\mathcal{P}$ by $v_0, v_1, ..., v_{n-1}$ in the order they appear along the boundary starting at v_0. For each index i, the two vertices v_i and v_{i+1} are *neighbors* (along the boundary). We write $\mathrm{chain}(v_i, v_j)$ to denote the sequence of vertices $v_i, v_{i+1}, \ldots, v_{j-1}, v_j$. For two points x, y on the boundary of $\mathcal{P}$ let $\widetilde{xy}$ be the part of the boundary of $\mathcal{P}$ between x and y in boundary order. If $\widetilde{xy}$ does not contain vertices of $\mathcal{P}$, we define $\mathrm{chain}(x, y)$ to be empty. Otherwise we define $\mathrm{chain}(x, y) := \mathrm{chain}(u, w)$, where u, w are the first and last vertices of $\mathcal{P}$ on $\widetilde{xy}$ in boundary order, respectively. In many cases only an upper bound $\bar{n} \geq n$ on the size of a polygon is known. Here and throughout, all operations on indices of vertices are understood to be modulo n or $\bar{n}$, depending on the context.

$\mathrm{chain}(v_i, v_j)$

Definition 2.9. Let u, v be two distinct vertices of a polygon $\mathcal{P}$ with $\overline{uv} \subset \mathcal{P}$. We say that u and v *see each other*, and refer to $\overline{uv}$ as the *line of sight* between u and v.

visibility

Consider Figure 2.1 along with the following definitions.

Definition 2.10. Let u, v be two vertices of a polygon $\mathcal{P}$ that see each other, and x be the point at which the ray $\overrightarrow{uv}$ first crosses the boundary of $\mathcal{P}$. We define the *right extension* of $\mathrm{chain}(u, v)$ to be $\mathrm{RE}(u, v) := \mathrm{chain}(u, x) \setminus \mathrm{chain}(u, v)$. Similarly, we define the *left extension* of $\mathrm{chain}(v, u)$ to be $\mathrm{LE}(v, u) := \mathrm{chain}(x, u) \setminus \mathrm{chain}(v, u)$.

RE / LE

REP / LEP

Definition 2.11. Let u, v be two vertices of a polygon $\mathcal{P}$ that see each other. We define the *right-extended pocket* $\text{REP}(u, v) := \text{chain}(u, v) \cup \text{RE}(u, v)$ and the *left-extended pocket* $\text{LEP}(u, v) := \text{chain}(u, v) \cup \text{LE}(u, v)$.

Proposition 2.12. *Let u, v be two vertices of a polygon $\mathcal{P}$ that are not neighbors and see each other. There exist two unique polygons $\mathcal{P}_1, \mathcal{P}_2$, with $\mathcal{P}_1 \cap \mathcal{P}_2 = \overline{uv}$ and $\mathcal{P}_1 \cup \mathcal{P}_2 = \mathcal{P}$. We say that $\mathcal{P}_1$ and $\mathcal{P}_2$ are obtained by* cutting $\mathcal{P}$ along $\overline{uv}$.

Proof. By Proposition 2.8, we can split the boundary of $\mathcal{P}$ at u and v and close each of the two resulting polygonal curves using the line segment $\overline{uv}$. The resulting closed polygonal curves do not self-intersect, because no other vertex can lie on $\overline{uv}$. By Proposition 2.8, each of the new curves is the boundary of a polygon. We denote these two polygons by $\mathcal{P}_1$ and $\mathcal{P}_2$. Because the original boundary of the polygon did not self-intersect, and by definition of $\overline{uv}$, we have $\mathcal{P}_1 \cap \mathcal{P}_2 = \overline{uv}$. By definition, $\mathcal{P}_1 \cup \mathcal{P}_2 = \mathcal{P}$. □

triangulation

Definition 2.13. A *triangulation* of a polygon $\mathcal{P}$ is a set of triangles T such that

1. the union of all vertices of triangles in T is equal to the set of vertices of $\mathcal{P}$,
2. the interiors of any two triangles in T do not intersect,
3. $\mathcal{P}$ is the union of all triangles in T.

Proposition 2.14. *The edges of any triangle in a triangulation of a polygon $\mathcal{P}$ are lines of sight in $\mathcal{P}$.*

Proof. The vertices of a triangle T in the triangulation of a polygon $\mathcal{P}$ are vertices of $\mathcal{P}$. Therefore, its edges are line segments connecting vertices of $\mathcal{P}$ and, since $T \subseteq \mathcal{P}$, they lie in $\mathcal{P}$. □

Proposition 2.15. *Every polygon admits a triangulation.*

Proof. The claim follows immediately from Definition 2.4. □

Proposition 2.16. *Let $\mathcal{P}$ be a polygon and s be a line of sight between two vertices in $\mathcal{P}$. The following holds:*

1. *If s is an edge of $\mathcal{P}$, every triangulation of $\mathcal{P}$ has exactly one triangle with edge s.*
2. *Else, $\mathcal{P}$ admits a triangulation in which exactly two triangles have s as an edge.*

Proof. In order to prove the first statement we consider an edge s of $\mathcal{P}$ and any fixed triangulation T. Certainly, s cannot intersect the interior of any triangle in T. On the other hand, since T must cover s, there must be a triangle containing s as an edge. Since the interiors of two triangles may not overlap, there can only be one such triangle in T.

Now consider the case that s is not an edge of $\mathcal{P}$. Let u, v be the endpoints of s. If we cut $\mathcal{P}$ along $\overline{uv}$, we obtain two smaller polygons $\mathcal{P}_1$ and $\mathcal{P}_2$ (cf. Proposition 2.12). By Proposition 2.15 and the first part of the proof, there is a pair of triangulations of $\mathcal{P}_1$ and $\mathcal{P}_2$, respectively, that both contain exactly one triangle that has s as an edge. The union of both these triangulations is a triangulation of $\mathcal{P}$ which has exactly two triangles having s as an edge. □

Definition 2.17. An *ear* of a polygon $\mathcal{P}$ is a vertex v_i for which v_{i-1} and v_{i+1} see each other.

ear

Proposition 2.18. *Every polygon of size $n \geq 4$ has two ears that are not neighbors along the boundary.*

Proof. We prove this by induction on the size of the polygon. In a triangle, every vertex is an ear, since its two neighbors along the boundary are neighbors of each other. By definition, a polygon of size $n = 4$ is the union of two triangles that intersect along a line segment s. Let u, w be the vertices that are not endpoints of s. Both u and v are ears, since s is a line of sight in the polygon.

For the induction step, consider a polygon $\mathcal{P}$ of size $n > 4$. By Definition 2.4, there is a line of sight s which is not an edge of $\mathcal{P}$. Consider the two polygons $\mathcal{P}_1, \mathcal{P}_2$ obtained by cutting $\mathcal{P}$ along s. For $i \in [2]$, we claim that $\mathcal{P}_i$ has an ear u_i that is not an endpoint of s. The claim follows immediately if $\mathcal{P}_i$ is a triangle. Otherwise, by induction, $\mathcal{P}_i$ has two ears that are not neighbors along the boundary, hence one of these ears cannot be an endpoint of s. Now, since u_1,u_2 are no endpoints of s, they must be ears of $\mathcal{P}$.

By construction, u_1 and u_2 are not neighbors along the boundary of $\mathcal{P}$. □

Proposition 2.19. *Every polygon has two ears.*

Proof. The claim follows immediately from Proposition 2.18 and the fact that every triangle has three ears. □

Proposition 2.20. *Let v be an ear of a polygon $\mathcal{P}$ of size n. Cutting $\mathcal{P}$ along the line of sight between v's neighbors yields a triangle and a polygon $\mathcal{P}'$ of size $n-1$. We say $\mathcal{P}'$ is the polygon resulting from* cutting v off of $\mathcal{P}$.

Proof. The claim follows from the proof of Proposition 2.12. □

For three points x, y, z in the plane we use $\measuredangle_x(y, z)$ to denote the angle at x formed by the rays $\overrightarrow{xy}$ and $\overrightarrow{xz}$ in this order. Angles are measured in the same rotational direction as the fixed boundary order that is assumed for polygons. We will call angles larger than π *reflex* and all other angles *convex*. The next definition should be clear intuitively.

angle in $\mathcal{P}$

Definition 2.21. Let $\mathcal{P}$ be a polygon and $\overline{vu}$, $\overline{vw}$ be two lines of sight in $\mathcal{P}$. The *angle between $\overline{vu}$ and $\overline{vw}$ in $\mathcal{P}$* is $\measuredangle_v(u, w)$ if $u \in \text{chain}(v, w)$, and $\measuredangle_v(w, u)$ otherwise.

interior angle

Definition 2.22. The *interior angle* of a vertex v_i of a polygon $\mathcal{P}$ is $\measuredangle_{v_i}(v_{i+1}, v_{i-1})$. If $\measuredangle_{v_i}(v_{i+1}, v_{i-1}) > \pi$, we say v_i is *reflex,* otherwise v_i is *convex.*

Euclidean shortest path

Definition 2.23. A *Euclidean shortest path* in $\mathcal{P}$ between two vertices u, v of $\mathcal{P}$ is a shortest curve in $\mathcal{P}$ with u and v as endpoints.

The following theorem shows that there is a unique Euclidean shortest path and gives a characterization. We cite the result and do not give a proof here.

Theorem 2.24 ([36])**.** *The Euclidean shortest path $\mathcal{S}$ in a polygon $\mathcal{P}$ between two vertices u, v is a uniquely defined polygonal curve. Every interior vertex of $\mathcal{S}$ is a vertex of $\mathcal{P}$, and the angle in $\mathcal{P}$ between any pair of consecutive line segments in $\mathcal{S}$ is reflex.*

Proposition 2.25. *Let u, v be two vertices of a polygon $\mathcal{P}$ that see each other, and $w \in \mathrm{RE}(u,v) \cup \mathrm{LE}(v,u)$. Then, v is an interior vertex of the Euclidean shortest path from u to w.*

Proof. Let x be the point at which the ray $\overrightarrow{uv}$ first crosses the boundary of $\mathcal{P}$. By definition of RE and LE, any curve in $\mathcal{P}$ with endpoints u and w needs to cross $\overline{vx}$ at some point y. The claim now follows from the fact that the shortest curve from u to y consists of $\overline{uv}$ and $\overline{vy}$. □

2.3. Visibility Graphs

We start by briefly giving some usual definitions concerning general graphs. A *directed graph* $G = (V, A)$ is a pair of sets, where V contains the *vertices* of the graph and $A \subseteq V \times V$ contains its *arcs*. An arc $a = (u, v) \in A$ is an ordered pair of vertices $u, v \in V, u \neq v$, where u is the *source* of a and v is the *target* of a. We write $u = \mathrm{source}(a)$ and $v = \mathrm{target}(a)$ and say that v is *adjacent* to u, and (u, v) is *an arc at u*. If $(u, v) \in A$ and $(v, u) \in A$, we say $\{u, v\}$ is an *edge* of G. The *neighborhood* $\Gamma(u)$ of a vertex $u \in V$ is the set of vertices adjacent to u, i.e., $\Gamma(u) = \{x \in V | \, (u, x) \in A\}$, and the *degree* $d(u)$ of u is given by $d(u) = |\Gamma(v)|$. An *arc-labeled* directed graph $G = (V, A, \lambda)$ is a triple consisting of the set of vertices V and the set of arcs A of a directed graph, as well as a function λ that maps each arc a to its *arc-label* $\lambda(a)$, which can be any kind of object. A directed *multigraph* $G = (V, A)$ is a directed graph for which A is a multiset (i.e., A can contain the same arc multiple times) and for which arcs may have the same source and target. In order to correctly define arc-labeled multigraphs, we would need to extend arcs to consist of two vertices and a third parameter used to distinguish multiple copies of the same arc. For the sake of presentation, we abuse notation and write arcs of an arc-labeled multigraph as tuples, implicitly assuming that the arc-label function can distinguish between identical arcs using some hidden parameter.

The subgraph of a graph $G = (V, A)$ *induced* by a set of vertices $V' \subseteq V$ is the graph $G' = (V', A')$, with $A' = A \cap (V' \times V')$. Consider a sequence $w = (u_1, u_2, \ldots, u_k)$ of vertices in a directed graph $G = (V, A)$. Then w is a *walk* in G if $(u_i, u_{i+1}) \in A$ for each $i \in [k-1]$. The *length* of w is defined to be $k-1$. We say that u_1, u_k are the *source* and *target* of w, respectively, and denote them by $u_1 = \text{source}(w)$ and $u_k = \text{target}(w)$. All other vertices of w are *interior vertices* of w. We say w is a *path* if it is a walk and contains every vertex at most once. We say w is a *cycle* if $(u_1, u_2, \ldots, u_{k-1})$ is a path and $u_k = u_1$. Finally, w is a *Hamiltonian cycle* if it is a cycle and contains every vertex in V. If for every two vertices $u, v \in V$ there is a path in G from u to v, we say G is *strongly connected.* If $G = (V, A, \lambda)$ is an arc-labeled graph and w is a walk in G, we say that $\lambda(w) := (\lambda((u_1, u_2)), \lambda((u_2, u_3)), \ldots)$ is a *label-sequence* of G, or, more precisely, the *label-sequence associated with* w *in* G.

local orientation

Definition 2.26. A directed, arc-labeled graph $G = (V, A, \lambda)$ is *locally oriented* if every two arcs emanating from the same vertex have different labels.

Proposition 2.27. *Let* $G = (V, A, \lambda)$ *be a locally oriented graph, let* $v \in V$*, and let* Λ *be a sequence. There is at most one walk* w *in* G *with* $\text{source}(w) = v$ *and* $\lambda(w) = \Lambda$. *We define* $\Lambda(v) := w$ *if such a walk exists, and* $\Lambda(v) := \emptyset$ *otherwise. Similarly,* $\Lambda(G)$ *denotes the set of all walks in* G *with associated label-sequence* Λ.

Proof. We prove the claim by induction on the length of Λ. Let Λ_1 denote the first label in Λ and let Λ_{rest} denote the sequence without Λ_1. Because G is locally oriented, there is at most one arc $a_1 = (v, u)$ at v with $\lambda(a_1) = \Lambda_1$. Every walk w as in the claim must begin with this arc. If $|\Lambda| = 1$, we either have $w = (v, u)$ or there is no such arc. Now assume $|\Lambda| > 1$ and assume that the claim holds for every shorter label-sequence. If there is no walk w' with $\text{source}(w') = u$ and $\lambda(w') = \Lambda_{\text{rest}}$, there can also not be a walk w as in the claim. Otherwise, by induction, we have that there is a unique such walk w'. Then, the walk $w = v \oplus w'$ is the unique walk with $\lambda(w) = \Lambda$. □

From now on we will simply use the term "graph" to refer to directed, strongly connected multigraphs. We use the term "arc-labeled graph" as a shorthand for a directed, strongly connected, locally oriented, arc-labeled multigraph. We will occasionally explicitly state attributes redundantly for emphasis. By $\mathscr{G}$ we denote the family of all arc-labeled graphs.

Definition 2.28. Let $\mathcal{P}$ be a polygon. The (unlabeled) *visibility graph* G_{vis} of $\mathcal{P}$ is a graph (V, A) where V is the set of vertices of $\mathcal{P}$ and A consists of all ordered pairs of vertices that see each other in $\mathcal{P}$.

visibility graph

Let $\mathcal{P}$ be a polygon and G_{vis} be its visibility graph. Every edge of G_{vis} corresponds to a line of sight in $\mathcal{P}$ and we use both terms interchangeably speaking of an angle between two edges of G_{vis} when we mean the angle between the corresponding lines of sight in $\mathcal{P}$, etc. In order to avoid confusion between edges of $\mathcal{P}$ and edges of G_{vis}, we from now on refer to the edges of $\mathcal{P}$ as its *boundary edges.* For a vertex v_i of $\mathcal{P}$, we define $\text{vis}(v_i) := \text{chain}(v_{i+1}, v_{i-1}) \cap \Gamma(v_i)$ to denote the sequence of vertices visible to v_i in boundary order. Accordingly, $\text{vis}_l(v_i)$ is the l-th vertex visible to v_i in boundary order starting at v_i. Conversely, $O_{v_i}(v_j)$ is used to denote the index x such that $\text{vis}_x(v_i) = v_j$.

Definition 2.29. Let $\mathcal{P}$ be a polygon and $G_{\text{vis}} = (V, A)$ be its visibility graph. The arc-labeled graph (V, A, λ) is *an arc-labeled visibility graph* of $\mathcal{P}$ if there is a mapping φ with $\varphi(\lambda(a)) = O_u(w)$ for every arc $a = (u, w)$ in A.

arc-labeled visibility graph

In other words, we require every arc (v_i, v_j) of an arc-labeled visibility graph G_{vis} to encode $O_{v_i}(v_j)$ in its label, i.e., the arcs at a vertex are ordered by the boundary order of their targets. It is easy to see that such an arc-labeling is a local orientation of G_{vis}. We will later encounter arc-labelings with more complex labels that encode additional information. By $\mathscr{F} \subseteq \mathscr{G}$ we denote the family of all arc-labeled visibility graphs.

Definition 2.30. A family $\mathscr{F}' \subseteq \mathscr{F}$ is *complete* if for every unlabeled visibility graph $G = (V, A)$ there is exactly one function λ such that $(V, A, \lambda) \in \mathscr{F}'$.

complete family

Definition 2.31. Let G_{vis} be a visibility graph and $C = (u_1, u_2, \ldots, u_k)$ be a cycle in G_{vis} with source v_i. If $(u_1, u_2, \ldots, u_{k-1})$ is a subsequence of $\text{chain}(v_i, v_{i-1})$, we say C is an *ordered cycle* in G_{vis}.

ordered cycle

Proposition 2.32. *Let $\mathcal{P}$ be a polygon, G_{vis} be its visibility graph, and C be an ordered cycle in G_{vis}. The edges along C form a closed polygonal curve that is the boundary of a subpolygon $\mathcal{P}' \subseteq \mathcal{P}$. The subgraph of G_{vis} induced by the vertices in C is the visibility graph of $\mathcal{P}'$. We will refer to $\mathcal{P}'$ as the* polygon induced by C.

Proof. Every edge along C is a line of sight in $\mathcal{P}$, and because C is ordered, no two such lines of sight can cross. Therefore, the edges along C form a polygonal curve $\mathcal{C}$. By Proposition 2.8, $\mathcal{C}$ is the boundary of a polygon $\mathcal{P}'$. Since $\mathcal{C} \subset \mathcal{P}$, it follows by Theorem 2.7 that $\mathcal{P}' \subseteq \mathcal{P}$. It remains to show that the visibility graph $G'_{\text{vis}} = (V', A')$ of $\mathcal{P}'$ is equal to the subgraph $G'' = (V'', A'')$ of G_{vis} induced by the vertices of C. Both V' and V'' are the set of vertices of C, hence $V' = V''$.

For every $a \in A'$ we have that a is a line of sight of $\mathcal{P}'$. Since $\mathcal{P}' \subseteq \mathcal{P}$, a must also be a line of sight of $\mathcal{P}$ and hence $a \in A''$. Now consider an arc $a \in A''$. If a is a boundary edge of $\mathcal{P}$, its two endpoints must be neighbors along the boundary of $\mathcal{P}'$, and we thus have $a \in A'$. Otherwise, since both endpoints of a are on C, and since C is ordered, the curve $\mathcal{C}$ cannot cross a. Assume, for the sake of contradiction, that $a \notin A'$. Then, $\mathcal{C}$ must be contained in one of the two polygons obtained by cutting $\mathcal{P}$ along a. Hence, C is an ordered cycle also in the visibility graph of this subpolygon. But in this subpolygon, a is a boundary edge which contradicts $a \notin A'$ as we saw before. □

Proposition 2.33. *Let v_i be a vertex of a polygon $\mathcal{P}$ of size $n > 3$. If $d(v_i) = 2$, then $d(v_{i-1}) > 2$ and $d(v_{i+1}) > 2$.*

Proof. By Proposition 2.16, every triangulation of $\mathcal{P}$ contains exactly one triangle using the edge $\overline{v_i v_{i-1}}$ (or $\overline{v_i v_{i+1}}$). Let u be the third vertex in one such triangle. Then v_i sees u and so does v_{i-1} (resp. v_{i+1}). From $d(v_i) = 2$ it follows that u must be a neighbor of v_i along the boundary, i.e., $u = v_{i+1}$ (resp. $u = v_{i-1}$). Because of $n > 3$, u cannot at the same time be a neighbor of v_{i-1} (resp. v_{i+1}). Because v_{i-1} (resp. v_{i+1}) sees u in addition to its two neighbors, we have $d(v_{i-1}) > 2$ (resp. $d(v_{i+1}) > 2$). □

Proposition 2.34. *No two consecutive vertices along an ordered cycle C of length $|C| > 3$ have degree two in the subgraph induced by the vertices of C.*

Proof. The claim follows from Propositions 2.32 and 2.33. □

Definition 2.35. Let $\mathcal{P}$ be a polygon and v_i, v_j be two vertices of $\mathcal{P}$ that do not see each other. A vertex $v_b \in \text{chain}(v_{i+1}, v_{j-1})$ is a *blocker* of (v_i, v_j) if no vertex in $\text{chain}(v_i, v_{b-1})$ sees a vertex in $\text{chain}(v_{b+1}, v_j)$.

blocker

Proposition 2.36. *Let v_i, v_j, v_a, v_b be four distinct vertices of a polygon $\mathcal{P}$ with $v_a \in \text{chain}(v_{i+1}, v_{j-1})$ and $v_b \in \text{chain}(v_{a+1}, v_{j-1})$. Then v_a and v_b both block (v_i, v_j) if and only if v_a blocks (v_i, v_b) and v_b blocks (v_a, v_j).*

Proof. First assume that v_a and v_b both block (v_i, v_j). By definition, no vertex in $\text{chain}(v_i, v_{a-1})$ can see a vertex in $\text{chain}(v_{a+1}, v_j)$ and hence in $\text{chain}(v_{a+1}, v_b)$, since $v_b \in \text{chain}(v_{a+1}, v_{j-1})$. In other words v_a blocks (v_i, v_b). Similarly, v_b blocks (v_a, v_j).

Now assume that v_a blocks (v_i, v_b) and v_b blocks (v_a, v_j). For the sake of contradiction assume that v_a (a similar argument holds for v_b) does not block (v_i, v_j). Then, there is a pair of vertices $v_x \in \text{chain}(v_i, v_{a-1})$, $v_y \in \text{chain}(v_{b+1}, v_j)$ that see each other. We choose v_x, v_y such that $|\text{chain}(v_x, v_y)|$ is minimal. Let v'_x be the last vertex in $\text{chain}(v_x, v_a)$ that is visible to v_x and v'_y be the first vertex in $\text{chain}(v_a, v_y)$ that is visible to v_y. Note that $v'_y \neq v_a$ since v_b blocks (v_a, v_j). Both v_x and v_y have degree two in the subgraph induced by the ordered cycle $(v_x) \oplus \text{chain}(v'_x, v'_y) \oplus (v_y)$. Also this ordered cycle has length greater three since $v'_x \neq v'_y$ by definition. The existence of such a cycle, however, is a contradiction to Proposition 2.34. □

Proposition 2.37. *Let v_i, v_j be two vertices of a polygon $\mathcal{P}$. Every interior vertex of the Euclidean shortest path from v_i to v_j is a blocker of (v_i, v_j) or of (v_j, v_i).*

Proof. We prove the claim by induction on the length of the Euclidean shortest path. For length one the claim is trivial, and for length two it follows from the fact that the two line segments of the shortest path form an angle larger than π inside $\mathcal{P}$ (Theorem 2.24). Assume the Euclidean shortest path $u_1 = v_i, u_2, u_3, \ldots, u_{k-1}, u_k = v_j$ has length $k-1 > 3$. We can, without loss of generality, prove the claim for any vertex u_l with $1 < l \leq \lceil k/2 \rceil$. It is clear that the Euclidean shortest path from v_i to u_{l+1} is $v_i, u_2, \ldots, u_l, u_{l+1}$ and the Euclidean shortest path from u_l to v_j is $u_l, u_{l+1}, \ldots, v_j$. Therefore, by induction, we know that u_l blocks (v_i, u_{l+1}) or (u_{l+1}, v_i) and u_{l+1} blocks (u_l, v_j) or (v_j, u_l). If u_l blocks (v_i, u_{l+1}) and u_{l+1} blocks (v_j, u_l), we have $v_j \in \text{chain}(u_l, u_{l+1})$, and hence u_l blocks (v_i, v_j) since it blocks (v_i, u_{l+1}). Similarly, if u_l blocks (u_{l+1}, v_i) and u_{l+1} blocks (u_l, v_j), we have that u_l blocks (v_j, v_i). If u_l blocks (v_i, u_{l+1}) and u_{l+1} blocks (u_l, v_j), then both u_l and u_{l+1} block (v_i, v_j) by Proposition 2.36. Finally, if u_l blocks (u_{l+1}, v_i) and u_{l+1} blocks (v_j, u_l), then, again by Proposition 2.36, both u_l and u_{l+1} block (v_j, v_i). □

We now have at our disposal a set of properties of visibility graphs that will become useful in later chapters. The full characterization of visibility graphs is a long standing open problem [1, 24, 23, 30, 31, 48], i.e., it is still unknown what properties exactly a graph needs to fulfill in order to be the visibility graph of a polygon. Currently, four different necessary conditions have been established in the literature, but there are still graphs that satisfy them all without being valid visibility graphs. Even though we do not investigate the characterization of visibility graphs in this thesis, we give the known conditions for the reader's convenience. We first need a bit more terminology.

minimal invisible pair

Definition 2.38. A *minimal invisible pair* is a pair of vertices v_i, v_j that do not see each other, such that both (v_i, v_j) and (v_j, v_i) have at most one blocker.

separability

Definition 2.39. Let $(v_i, v_j, v_l, v_k) \subseteq \text{chain}(v_{a+1}, v_{a-1})$ for some vertex v_a. Then the pairs v_i, v_j and v_l, v_k are called *separable* with respect to v_a.

blocker assignment

Definition 2.40. A *blocker assignment* is a mapping from minimal invisible pairs to vertices such that:

1. every minimal invisible pair v_i, v_j maps to the blocker of either (v_i, v_j) or (v_j, v_i),

2. if a minimal invisible pair v_i, v_j is mapped to a blocker v_b of (v_i, v_j), then every minimal invisible pair (v'_i, v'_j) with $v'_i \in \text{chain}(v_i, v_{b-1}), v'_j \in \text{chain}(v_{b+1}, v_j)$ is mapped to v_b,
3. no two minimal invisible pairs separable with respect to a vertex v_a are mapped to v_a.

We are now prepared to report the currently strongest known set of necessary conditions. For a detailed discussion with proofs, refer to [30].

Theorem 2.41 ([23, 29, 48]). *Every visibility graph fulfills the following conditions:*

1. *every ordered cycle of length $k \geq 4$ induces a subgraph with at least $2k - 3$ edges,*
2. *for every pair of vertices v_i, v_j that do not see each other, there is a blocker of either (v_i, v_j) or (v_j, v_i),*
3. *there is a blocker assignment,*
4. *for any ordered cycle D, the number of different vertices in D assigned to minimal invisible pairs of vertices in D is at most $|D| - 3$.*

As mentioned above, a graph is known which fulfills all necessary conditions without being a visibility graph [49] (cf. Figure 2.2).

2.4. The Agent Model

An agent exploring an arc-labeled graph $G = (V, A, \lambda)$ is an entity that moves from vertex to vertex along arcs of G. More precisely, we define the agent model by making the following assumptions:

1. The agent is at all times located at some (not necessarily the same) vertex of G.
2. The agent has unlimited memory.
3. The agent can perform any kind of computation on the data it has stored.
4. The only information about G that the agent can access is the set of arc-labels $\mathcal{L}_G(v) := \{\lambda(v, u) | (v, u) \in A\}$ of the arcs at its current location v.

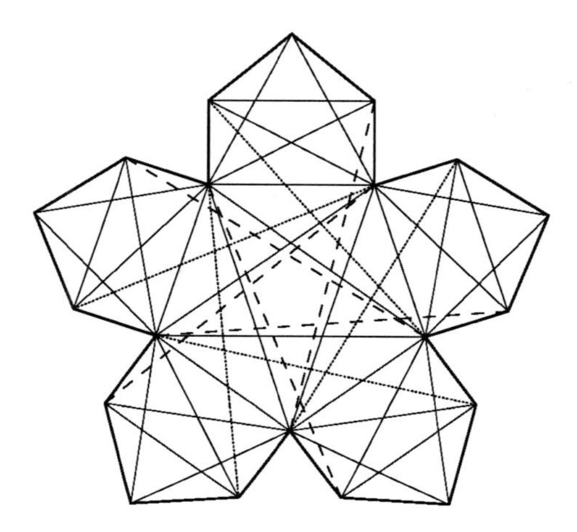

Figure 2.2.: Example of a graph that fulfills all known necessary conditions for visibility graphs without being one.

5. The agent can select an arc-label $L \in \mathcal{L}_G(v)$ among the arc-labels at its current location v, and move (instantaneously) to the target of the corresponding arc. We say the agent *moves according to L*.

Note that the agent can distinguish the arcs at its location only by their labels. In particular, the agent cannot distinguish the arc that leads to its previous location. We emphasize that an agent exploring an arc-labeled graph G has no access to global vertex identities. In fact, we are interested in the *graph reconstruction problem* which is defined as

"Find an arc-labeled graph isomorphic to G."

More precisely, we investigate different families of graphs with respect to the question whether the graph reconstruction problem can always be solved by an agent.

Definition 2.42. An *exploration strategy* is a terminating algorithm that governs the movements and computations of an agent, depending only on the information gathered during the execution of the algorithm.

The time spent by an agent executing an exploration strategy $\mathfrak{A}$ is the total number of moves and computational steps performed by

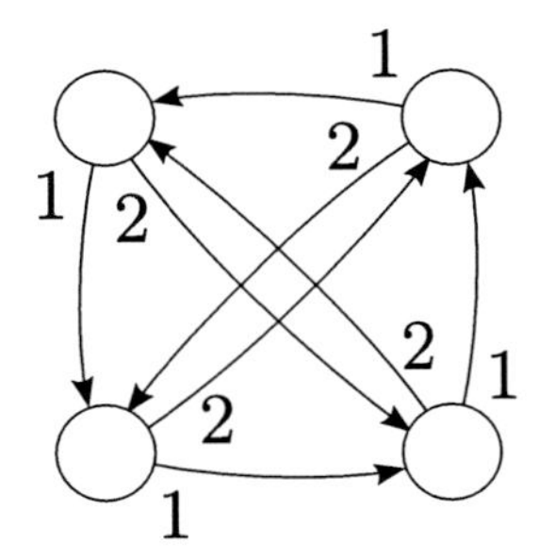

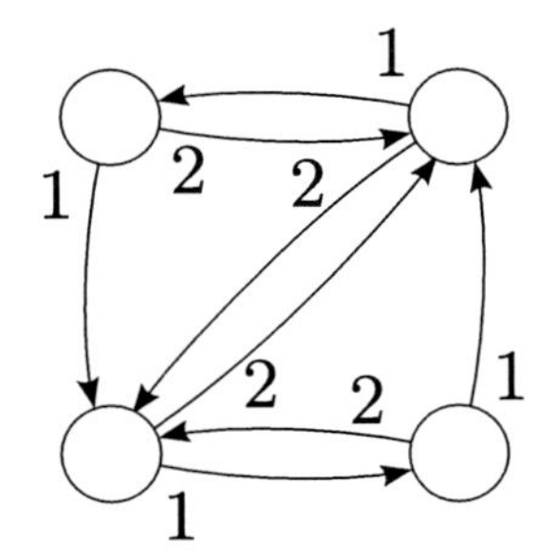

Figure 2.3.: Two non-isomorphic graphs that cannot be distinguished by an agent.

the agent upon termination of $\mathfrak{A}$. We say that $\mathfrak{A}$ is *polynomial* for a family $\mathscr{G}' \subseteq \mathscr{G}$ of arc-labeled graphs if there is a polynomial $p(n)$ such that for an agent exploring any graph of $G \in \mathscr{G}'$, executing $\mathfrak{A}$ takes time at most $p(n)$, where n is the size of G. We say that $\mathfrak{A}$ computes some quantity in $\mathscr{G}'$ if an agent exploring any graph in $\mathscr{G}'$ computes the quantity during the execution of $\mathfrak{A}$. Consider an agent exploring a fixed graph $G \in \mathscr{G}'$. We say that $\mathfrak{A}$ moves the agent to vertex v of G if the agent is located at v upon termination of $\mathfrak{A}$. If $\mathfrak{A}$ always moves an agent exploring any graph in $\mathscr{G}'$ to its initial location, we say $\mathfrak{A}$ is *returning.*

Definition 2.43. Let $\mathscr{G}' \subseteq \mathscr{G}$ be a family of arc-labeled graphs. We say an agent *can solve the reconstruction problem* in $\mathscr{G}'$ if there is an exploration strategy that, when executed while exploring any graph $G \in \mathscr{G}$, computes a graph isomorphic to G.

For example, on the class of arbitrary graphs and labelings, the agent cannot solve the reconstruction problem. Intuitively, this is because there are non-isomorphic graphs that are "indistinguishable" to the agent, i.e., any observations made in one of them by the agent could originate from both (cf. Figure 2.3). We will now capture this intuition formally.

indistinguishable vertices

Definition 2.44. Let $G = (V, A, \lambda)$, $G' = (V', A', \lambda')$ be two arc-labeled graphs and $v \in V, v' \in V'$. If every exploration strategy computes the same result when executed by an agent with initial location v and an agent with initial location v', we say v and v' are *indistinguishable.*

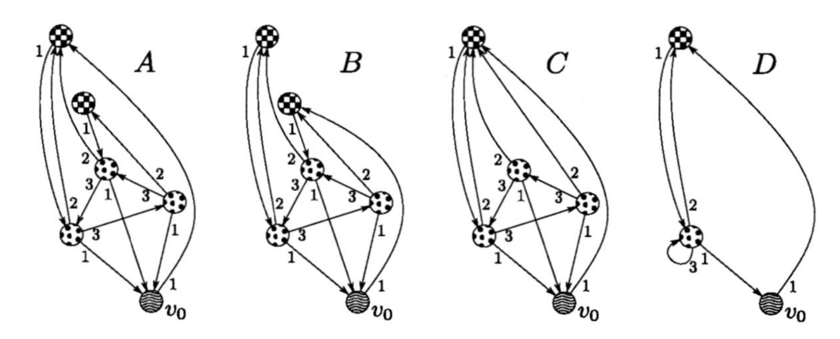

Figure 2.4.: Four graphs that are indistinguishable to an agent initially located at vertex v_0 in terms of observations made by the agent along any walk. Vertices which appear indistinguishable to the agent can be merged in order to obtain smaller but still indistinguishable graphs. Graph D is prime and hence the minimum base graph of all four graphs.

indistinguishable graphs

Definition 2.45. Two arc-labeled graphs $G = (V, A, \lambda)$, $G = (V', A', \lambda')$ are said to be *indistinguishable* if there are indistinguishable vertices $v \in V, v' \in V'$.

The following proposition serves as a motivation for the definition of indistinguishability. We defer the proof to Chapter 7.

Proposition 2.46. *An agent can solve the reconstruction problem in a family $\mathscr{G}' \subseteq \mathscr{G}$ if and only if there are no two non-isomorphic, indistinguishable arc-labeled graphs in $\mathscr{G}'$.*

Proof. The "if" follows from the definition of indistinguishable graphs. The inverse follows from Theorem 7.3. □

reconstructing a graph

Definition 2.47. Let $\mathscr{G}' \subseteq \mathscr{G}$ be a family of arc-labeled graphs. If $G \in \mathscr{G}'$ is not indistinguishable from any other non-isomorphic graph in $\mathscr{G}'$, we say G *can be reconstructed* in $\mathscr{G}'$.

Consult Figure 2.4 along with the formal statements below.

Proposition 2.48. *Let $G = (V, A, \lambda)$, $G' = (V', A', \lambda')$ be two graphs and $v \in V, v' \in V'$. The vertices v and v' are indis-*

tinguishable if and only if, for every label-sequence Λ*, we have* $\Lambda(v) \neq \emptyset \Leftrightarrow \Lambda(v') \neq \emptyset$.

Proof. Assume v and v' are indistinguishable and, for the sake of contradiction, that there is a label-sequence Λ with $\Lambda(v) \neq \emptyset$ and $\Lambda(v') = \emptyset$, or vice versa. Then, moving according to the labels in Λ is possible when starting from exactly one of the two vertices. The exploration strategy that simply detects this difference contradicts the assumption that v and v' are indistinguishable.

Conversely, assume that for every label-sequence Λ we have $\Lambda(v) \neq \emptyset \Leftrightarrow \Lambda(v') \neq \emptyset$. Then any exploration strategy the agent executes gets the same input for initial positions v and v'. Therefore all its computations need to yield the same result. □

Proposition 2.49. *Let* $G = (V, A, \lambda), G' = (V', A', \lambda')$ *be two indistinguishable arc-labeled graphs. For every vertex* $v \in V$ *there is a vertex* $v' \in V'$ *such that* v *and* v' *are indistinguishable.*

Proof. Consider any fixed vertex $v \in V$. By assumption, there are two indistinguishable vertices $u \in V, u' \in V'$. Since G is strongly connected, there is a path p from u to v. We set $\bar{\Lambda} := \lambda(p)$ and have that $\bar{\Lambda}(u') \neq \emptyset$, since u and u' are indistinguishable. We claim that $v' := \text{target}(\bar{\Lambda}(u'))$ and v are indistinguishable as well. Otherwise, there would be a label-sequence δ with $\delta(v) \neq \emptyset$ and $\delta(v') = \emptyset$ or vice versa. With $\Lambda := \bar{\Lambda} \oplus \delta$, we would then have $\Lambda(u) \neq \emptyset$ and $\Lambda(u') = \emptyset$ or vice versa, which would be a contradiction to the fact that u and u' are indistinguishable. □

Theorem 2.50 ([8]). *Let* G *be an arc-labeled graph. There is a unique smallest graph amongst the graphs indistinguishable from* G *up to isomorphism.*

Definition 2.51. The *minimum base graph* $G^\star$ of an arc-labeled graph G is the smallest graph indistinguishable from G up to isomorphism.

minimum base graph

Definition 2.52. A *prime graph* is an arc-labeled graph G for which $G^\star$ is isomorphic to G.

prime graph

Proposition 2.53. *No two vertices of a prime graph are indistinguishable.*

Proof. Assume there was a prime graph $G = (V, A, \lambda)$ with two indistinguishable vertices $u, v \in V$. Consider the subgraph $G' = (V', A', \lambda)$ induced by $V' = V \setminus \{u\}$. Let $G'' = (V', A'', \lambda)$ with $A'' := A' \cup \{(w, v) | (w, u) \in A \wedge w \neq u\}$. It is easy to see that G'' is indistinguishable from G, which is a contradiction since G'' is smaller than G. □

classes

Definition 2.54. Let G be an arc-labeled graph. Consider the equivalence classes formed by indistinguishable vertices of G. We simply call them *classes* of G.

For a graph G and vertex v we write C_v to denote the class containing vertex v. By $v^\star$ we denote the vertex of $G^\star$ indistinguishable from v, and we set $C_{v^\star} := C_v$.

Proposition 2.55. *Let $G = (V, A, \lambda)$ be an arc-labeled graph and $a = (u, w) \in A$. Then, for every $v \in C_u$, there is exactly one an arc b at v with $\lambda(b) = \lambda(a)$. We have $C_{\text{target}(b)} = C_{\text{target}(a)}$.*

Proof. Let u be any fixed vertex of C_v. By definition, v and u are indistinguishable, and therefore v must have an arc b with label $\lambda(a)$. Because G is locally orientated, there can only be one such arc. The targets a and b must be indistinguishable in order for u and v to be indistinguishable. Therefore, $C_{\text{target}(b)} = C_{\text{target}(a)}$. □

In an arc-labeled visibility graph the arc-label of an arc (u, w) encodes $O_u(w)$. Proposition 2.55 allows us to write $C_u(O_u(w)) := C_w$. By the same token, every vertex in C_u has the same degree, which we will denote by $d(C_u)$.

Proposition 2.56. *Let $\mathcal{B}$ be the sequence of classes along the boundary in an arc-labeled visibility graph $G_{\text{vis}} = (V, A, \lambda)$. Then, there is an integer k and a sequence $\mathcal{B}^\star$ containing every class of G_{vis} exactly once, such that $\mathcal{B} = \bigoplus_{i=1}^{k} \mathcal{B}^\star$.*

Proof. Let $v \in V$ and $u \in C_v$. Consider the Hamiltonian cycles p, q that contain the vertices of G_{vis} in boundary order starting with v, u, respectively. Recall that λ encodes the boundary order of each arc. Every arc in both p and q must be the first one at its source in boundary order. Therefore $\lambda(p) = \lambda(q)$. By Proposition 2.55, the classes of the vertices along p must be the same as those along q. This holds for every pair of indistinguishable vertices, and both p and q visit every vertex. The claim follows. □

Proposition 2.57. *All classes of an arc-labeled visibility graph have equal size.*

Proof. The proof is immediate from Proposition 2.56. □

Proposition 2.58. *Let $G_{\text{vis}} = (V, A, \lambda)$ be a visibility graph with a vertex $v \in V$ that is not indistinguishable from any other vertex in V. Then G_{vis} is a prime graph.*

Proof. Because $G^\star$ has a vertex for every class of G, it is sufficient to show that every class of G has size one. By definition, this is the case for C_v. Hence every class has size one, by Proposition 2.57. □

Proposition 2.59. *Let $\mathscr{G}_n$ be the family of arc-labeled graphs of size n. Every prime graph $G \in \mathscr{G}_n$ can be reconstructed in $\mathscr{G}_n$.*

Proof. We need to show that no graph G' non-isomorphic to G is indistinguishable from G. This follows immediately from Theorem 2.50 and Definition 2.52. □

Together, the last two propositions state that a visibility graph G_{vis} can always be reconstructed if at least one vertex $v^\star$ of it can be distinguished from all other vertices. The derivation of this insight was somewhat technical, but Figure 2.5 illustrates intuitively how an agent can easily reconstruct G up to isomorphism (assuming $v_0 = v^\star$).

2.4.1. Agents Exploring Polygons

In this thesis we are interested in simplistic agents that explore a polygonal environment trying to draw a map. We model this scenario as the exploration of a visibility graph by an agent trying to solve the graph reconstruction problem. This means that, by definition, the agent moves between vertices of the polygon

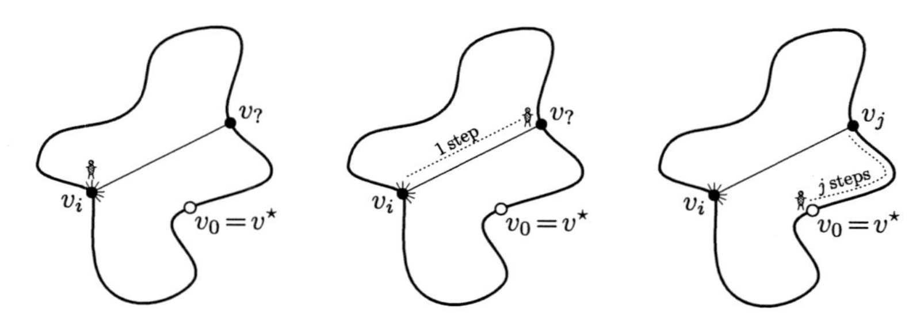

Figure 2.5.: If the agent can distinguish some vertex $v^\star$ (w.l.o.g. $v^\star = v_0$), it can easily determine where edges lead: starting at a vertex v_i, the agent can identify the target of an edge by moving along this edge and then along the boundary until it encounters $v^\star$, counting the number of moves it makes.

along lines of sight. While located at a vertex v of the polygon, the agent "sees" all other vertices visible to v and is able to order them according to their order along the boundary, starting and ending with v's neighbors. The ability to order the visible vertices is reflected by the fact that the arc-labeling of an arc-labeled visibility graph is defined to encode the order of the arcs emanating from each vertex.

In later chapters we will consider various extensions of the agent's capabilities. There are different kinds of extensions that we can make. For example, we might allow the agent to measure certain angles inside the polygon, or we might assume that the agent knows the total number of vertices already beforehand. In order to maintain our perspective of an agent exploring an arc-labeled visibility graph G_{vis}, we will encode the additional information accessible by the agent in the arc-labeling of G_{vis}. This will define a complete family of arc-labeled visibility graphs. As formulated in Proposition 2.46, the agent can solve the reconstruction problem, and thus map any polygon, if and only if no two graphs of this family are indistinguishable.

encoding agent models

Definition 2.60. Let $\mathscr{F}' \subset \mathscr{F}$ be a family of arc-labeled visibility graphs and assume a fixed agent model. We say $\mathscr{F}'$ *encodes* the agent model if for every polygon $\mathcal{P}$, there is an arc-labeled

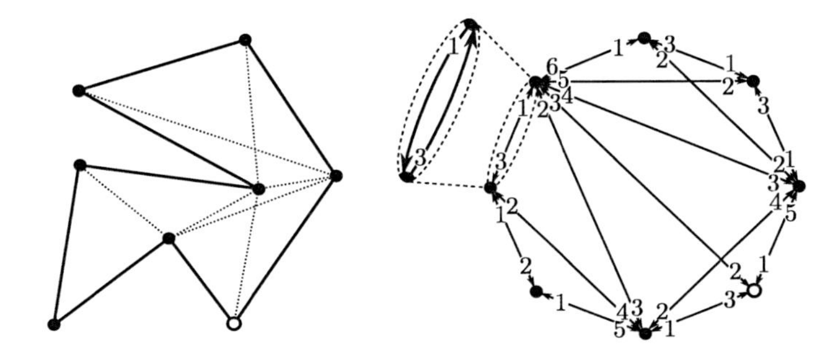

Figure 2.6.: A polygon with the corresponding directed and arc-labeled visibility graph. Every bidirected edge represents two arcs of opposite orientation.

graph $G_{\text{vis}} \in \mathscr{F}'$ which is a visibility graph of $\mathcal{P}$ and encodes the observations that an agent can make at each vertex v of $\mathcal{P}$ in the arc-labels of the arcs at v in G_{vis}.

We will now describe how the individual extensions to the agent's capabilities can be encoded in the arc-labeling of the visibility graph. We can form any combinations of these extensions simply by combining the corresponding arc-labels for every arc. Note that irrespective of the extensions made to the agent's capabilities, the arc-labeling of an arc-labeled visibility graph is, by definition, guaranteed to encode the order of the arcs at every vertex. This basic property can be achieved easily by labeling each arc by its position in the local order at its source (cf. Figure 2.6).

Among the most natural extensions to the agent's sensing is the addition of a distance or angle sensor. A distance sensor allows the agent to infer its Euclidean distance to every visible vertex, while an angle sensor allows it to measure the angle inside the polygon between any two arcs incident to its current location (cf. Figure 2.7). We use four different types of angle sensors (cf. Figure 2.8):

1. The *standard angle sensor* allows to measure every angle exactly.
2. The *angle-type sensor* only allows to distinguish whether an angle is greater π or not.

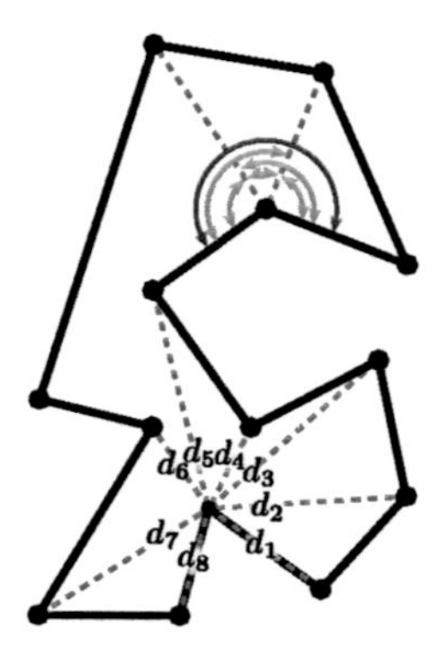

Figure 2.7.: Illustration of distance and angle sensors.

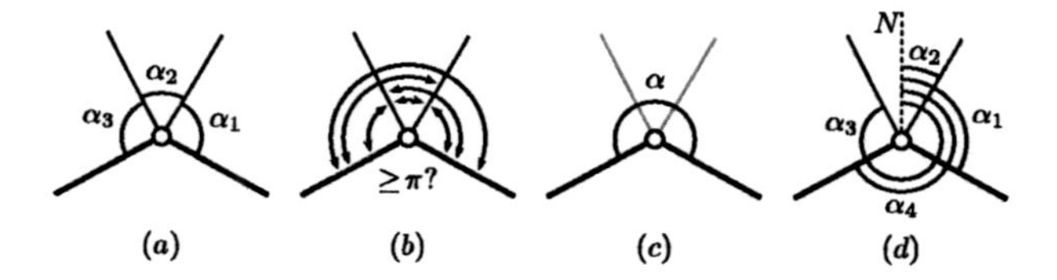

Figure 2.8.: Local perception with different kinds of angle sensors. From left to right: angle sensor, angle-type sensor, inner-angle sensor, compass with global reference N.

3. The *inner-angle sensor* allows only to measure the interior angle at the agent's location.
4. The *compass* defines a global reference direction and allows the agent to measure the angle of each arc with respect to this direction.

The data made accessible by each of these sensors can be encoded in the arc-labeling of the visibility graph in a similar fashion. As an example, consider the data available through the angle-type sensor. Let $a_1, a_2, \ldots, a_{d(v)}$ be the arcs at vertex v. We can simply extend the label of arc a_i by a sequence s of bits of length $d(v)$, where $s_j = 1$ if a_i and a_j form a reflex angle inside the polygon, and $s_j = 0$ otherwise (cf. Figure 2.9). For the distance sensor, we can simply extend the label of each arc by its length.

Another sensor that we will use to extend the basic capabili-

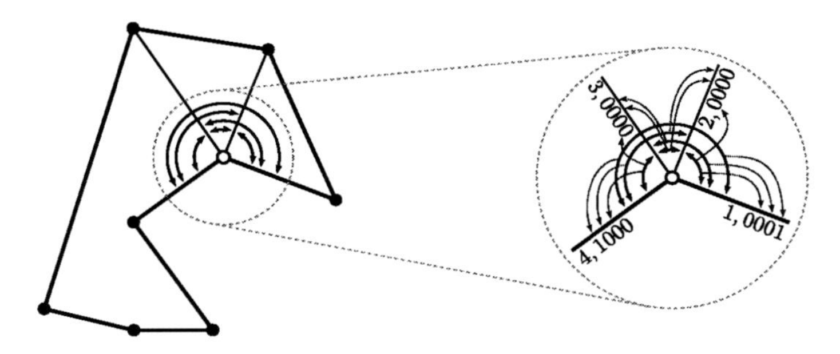

Figure 2.9.: Illustration of how to extend the basic arc-labeling for an angle-type sensor.

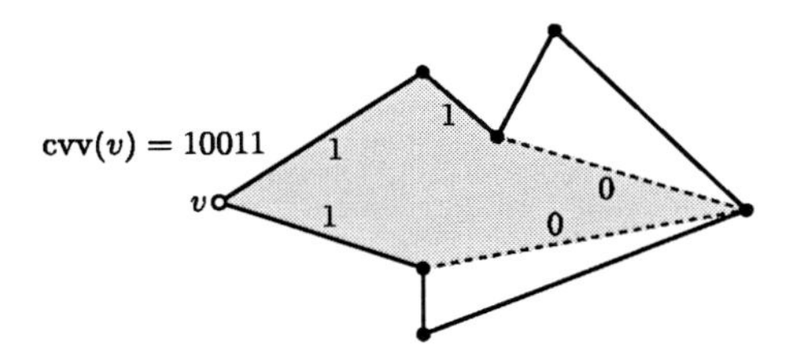

Figure 2.10.: The combinatorial visibility vector of a vertex v encodes which vertices visible to v are neighbors on the boundary.

ties of the agent is the *cvv sensor*. This sensor yields the *combinatorial visibility vector* of the agent's current location v: a sequence of bits $\text{cvv}(v) \in \{0,1\}^{d(v)+1}$ with the property that $\text{cvv}_1(v) = \text{cvv}_{d(v)}(v) = 1$, and $\text{cvv}_j(v) = 1$ for $1 < j \leq d(v)$ if and only if the $(j-1)$-th and j-th vertices that v sees (in boundary order) are neighbors on the boundary (cf. Figure 2.10). Intuitively, the cvv sensor provides information between which of the vertices visible to v there are other vertices which are not seen by v.

We mentioned before that the agent cannot distinguish the arc that leads to its last location – we say the agent cannot *look back*. In other words, the agent has no direct way of backtracking its moves. It is natural to consider an extended model that does not have this limitation. To this purpose we introduce a *look-back*

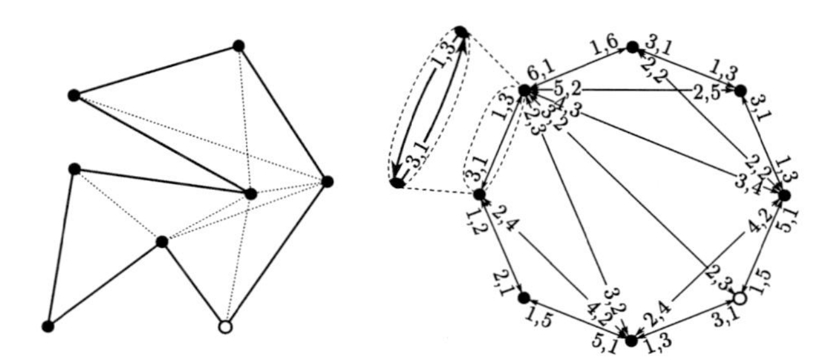

Figure 2.11.: An illustration of how the look-back capability can be encoded in the arc-labeling of the visibility graph from Figure 2.6.

sensor that provides the identity of the arc the agent would have to choose in order to backtrack its last move. Note that in a visibility graph every arc has an arc of opposite orientation. We can adapt the arc-labeling to reflect the look-back capability by adding to every arc-label the standard label of the opposing arc (cf. Figure 2.11).

Most of the time, we will assume the agent to be aware of the total number of vertices n or at least an upper bound $\bar{n} \geq n$. We say *the agent knows* n or *the agent knows* $\bar{n}$, respectively. We can see this assumption as another extension to the agent's capabilities, and reflect it in the arc-labeling by adding n (or $\bar{n}$) to every arc-label (technically, adding it to the label of one arc at each vertex would be sufficient).

2.4.2. Visibility Graph Reconstruction and Rendezvous

Throughout this thesis we are concerned with the *visibility graph reconstruction problem.* This problem is defined with respect to a fixed choice of set of extensions to the basic agent model. As described above, these extensions can be reflected in the arc-labeling of an arc-labeled visibility graph. We say that an agent *can solve the visibility graph reconstruction problem*, if it can solve the reconstruction problem for any family $\mathscr{F}' \subseteq \mathscr{F}$ that encodes the agent model. We say that the agent *can reconstruct an arc-labeled visibility graph* G_{vis} if for every family $\mathscr{F}' \subseteq \mathscr{F}$ that encodes the agent model there is no other graph in $\mathscr{F}'$ that is non-isomorphic to, and indistinguishable from G_{vis}.

Another very natural problem that becomes important when multiple agents need to cooperate is the *rendezvous problem*. We always assume agents to be deterministic, identical, and indistinguishable, in particular all agents execute the same deterministic exploration strategy. We only allow an agent to count the number of agents present at every vertex visible to it. We distinguish two variants of the problem. The *strong rendezvous problem* requires all agents to gather at the same vertex, while the *weak rendezvous problem* requires the agents to position themselves that they are all mutually visible to each other. We say that agents can *strongly/weakly meet* in an arc-labeled graph G if there is a deterministic exploration strategy $\mathscr{A}$ that, for every combination of starting locations, moves any number of agents executing $\mathscr{A}$ to positions that establish (strong or weak) rendezvous. We say that agents can *solve the strong/weak rendezvous problem* in a family of arc-labeled graphs $\mathscr{F}' \subseteq \mathscr{F}$ if there is a deterministic exploration strategy $\mathscr{A}$ that, for every graph in $\mathscr{F}'$ and every combination of starting locations, moves any number of agents to positions that establish (strong or weak) rendezvous.

Proposition 2.61. *If agents can strongly meet in an arc-labeled graph G, then G is a prime graph.*

Proof. Assume that G is not prime and let C be a class of G. We have $|C| > 1$. Position one agent on each vertex of C and assume the agents make movement decisions simultaneously. Because the vertices of C are indistinguishable, and because all agents execute the same exploration strategy, all agents make the same movement decisions and hence their locations always remain indistinguishable. Hence, the agents maintain their formation indefinitely. □

We will see in Section 7.3 that the converse holds as well.

2.4.3. Operations of the Agent

So far we have adopted an omniscient view on polygons and their visibility graphs. In order to express the local nature of an agent's perception when speaking about strategies for the agent, we now introduce some formalism for atomic operations of the agent. In the following, assume the agent is located at some vertex v and let $i, j \in [d(v)]$.

By '**degree**' we denote $d(v)$. By '**move to** i' we denote the operation that moves the agent along the i-th arc at v. By '**look back**' we denote the operation yielding the index $b \in [\mathrm{d}]$ such that $\mathrm{vis}_b(v)$ is the vertex the agent visited before v. Let $\angle_v(i, j) := \measuredangle_v(\mathrm{vis}_i(v), \mathrm{vis}_j(v))$ (note the difference between '$\measuredangle$' and '$\angle$'). By $\angle(i, j)$ we denote the operation yielding $\angle_v(i, j)$. By $\angle_{\mathrm{reflex}}(i, j)$ we denote the operation yielding 'true' if $\angle_v(i, j)$ is reflex and 'false' otherwise. By $\angle$ we denote the operation yielding $\angle_v(1, d(v))$. By $\angle(i)$ we denote the operation yielding the angle between the i-th arc at v and a global reference direction. Finally, by cvv_l, $l \in [d(v) + 1]$ we denote the operation yielding $\mathrm{cvv}_l(v)$.

Chapter 3.

Related Work

A variety of minimalistic agent models have been studied, focusing on different types of environments and objectives [3, 27, 34, 50]. Some attempts at defining hierarchical relationships between models have been made [9, 22, 40]. The basic agent model that we introduced in the previous chapter originates from [50] and was studied previously in [9, 28, 35, 50]. We now give a brief overview over the most prominent results concerning simple agents.

One of the first problems that was considered for mobile agents is the *art-gallery problem* which asks to place guards at some vertices of a polygon of size n, such that every point of the polygon is visible to at least one of the guards. The famous art-gallery theorem asserts that every polygon can be guarded in this way by placing at most $\lfloor n/3 \rfloor$ guards [13, 25]. While the theorem assumes full knowledge of the geometry of the polygon, Ganguli et al. considered the art-gallery problem in an initially unknown polygon, where the guards are autonomous mobile agents [27]. The guards in this study are allowed to move freely inside the polygon and to communicate over any distance as long as they see each other. Ganguli et al. showed that $\lfloor n/2 \rfloor$ such mobile guards can always self-deploy at vertices such that the polygon is guarded. This result raises the question whether the gap between $\lfloor n/2 \rfloor$ and $\lfloor n/3 \rfloor$ is inherent, due to the fact that the global geometry of the polygon is initially unknown to the agents. Suri et al. showed that this is not the case, and in fact $\lfloor n/3 \rfloor$ guards can self-deploy to guard the polygon [50]. The paper considered the basic agent model of Chapter 2 and additionally equipped each agent with a cvv sensor and a *pebble*. The agent can drop or pick up the pebble at its current location and distinguish vertices that hold a pebble. Note that the resulting agent model is weaker than the one employed by Ganguli et al. Suri et al. showed that their guards can compute the visibility graph of any polygon and thus also a triangulation of the polygon. As shown by Fisk, the triangulation can be colored with three colors [25], which is sufficient in order to conclude a bound of $\lfloor n/3 \rfloor$ guards for the art-gallery problem.

Another common problem is to determine the number of vertices n of a polygonal environment. Evidently, computing n is a trivial task for an agent with a pebble – such an agent can even infer n if the environment has holes [50]. Similarly, the task is easy when some vertex is locally distinguishable from all other vertices and an upper bound on n is known to the agent, where the upper bound is needed in order for the agent to find this distinguishable vertex. In this thesis, we show that, without prior knowledge about n, an agent can infer the size of a polygon in the following three cases: (a) the agent is restricted to moving along the boundary only and can perform angle measurements (Chapter 5), (b) the agent can move freely along edges of the visibility graph, can look back and has an angle-type sensor (Chapter 6), (c) the agent can move freely along edges of the visibility graph, has an angle-type sensor and has a compass (Chapter 6). An agent with cvv sensor and look-back sensor, on the other hand, cannot infer the size of a polygon in general [9]. While the sensing capabilities of the agent model introduced in Section 2.4.1 are very simple and hence inherently robust, we require that all visible vertices can be perceived. Komuravelli et al. [35] considered a faulty scenario in which it can happen that the agent perceives two distant vertices as a single *virtual* vertex (e.g., vertices that appear very close to each other). Komuravelli et al. studied whether an agent equipped with pebbles can infer the size of the polygon, showed that with a single pebble this is not possible, conjectured that two pebbles are also still insufficient, and showed that three pebbles allow computing the size of the polygon.

Yet another problem in polygonal environments is to count the number k of *target* points inside the polygon. Of course, the sensing model of the agent has to be extended in order to make it aware of the target points first. For example, at any vertex, the agent might be able to perceive a list of visible targets and a list of visible vertices, rather than visible vertices only. Gfeller et al. [28] showed that an agent with a pebble cannot approximate the number of points within a factor of $2 - \varepsilon$, for any $\varepsilon > 0$. Let a ρ-approximation of the number of points k refer to an upper bound z with $k \leq z \leq \rho k$. The results of Gfeller et al. imply that an agent knowing the *vertex-edge visibility graph* of the polygon, together with its initial position in it, can compute a 2-approximation of the number of points. The vertex-edge visibility graph is a bipartite graph with a node for every vertex and every boundary edge of the polygon, with an edge between a node corresponding to a boundary edge and a node corresponding to a vertex if at least one point on the boundary edge is visible to the vertex. Note that the vertex-edge visibility graph induces the visibility graph [42]. Komuravelli et al. later showed that the vertex-edge visibility graph is not needed to compute a 2-approximation [35] – knowing the visibility graph instead is already sufficient, at the cost of an exponential running time. The idea is simply to iterate over all vertex-edge visibility graphs that are compatible with the given visibility graph and run the 2-approximation of Gfeller et al. on each of those. The output is the smallest estimation of the number of points encountered in the process.

An important problem in unbounded, two-dimensional environments is the coordination of multiple agents. There is a large variety of studies concerned with this setting, and we can only mention a few examples here. The most prominent problem in this context is the *rendezvous* or *convergence problem* in which the agents need to meet in one point or at least converge to one point. Various agent models have been considered. Some assume all agents to act in a synchronized fashion, where others do not [37, 38]. Some agent models make restrictions on available memory or the range of the agent's vision [3]. Other studies are concerned with robustness issues in terms of measurement imprecisions [2, 14]. Instead of requiring the agents to gather at a common location, sometimes they are required to position themselves according to a given pattern [51].

3.1. Reconstructing Polygons from Data

One major challenge for an agent mapping an unknown environment is that it needs to collect the required data in a systematic fashion. Assume instead that we are given some data that was measured in some way or other in the environment. We can now ask wether we can construct a map or even the exact geometry of the environment from this data alone. This is a typical problem in computational geometry. We now list some prominent results for the reconstruction of polygons from measurement data.

3.1.1. Constructing a Consistent Polygon

There are two main variants of the general problem of reconstructing a polygon from measurement data. The first variant asks to find *some* polygon $\mathcal{P}'$ that is *consistent* with the data measured in the original polygon $\mathcal{P}$. Being consistent means that this data could have originated from a series of measurements in $\mathcal{P}'$. Studies that consider this variant of the reconstruction problem usually focus on the complexity of the problem rather than the question whether a unique solution exists.

The first reconstruction problem that was studied in this context asked to construct a polygon which is consistent with a given visibility graph. The problem is only known to be in PSPACE, its complexity is still open [23].

Jackson and Wismath studied the problem of reconstructing an orthogonal polygon from the "stabbing information" at all vertices [32] (cf. Figure 3.1 (left)). An orthogonal polygon is a polygon $\mathcal{P}$ for which every boundary edge is either horizontal or vertical. The orthogonal polygon is assumed to have no three vertices lying on a vertical or horizontal line, i.e., in particular, every vertex has a horizontal and a vertical boundary edge adjacent to it. A *horizontal stab* of vertex v is the horizontal ray starting at v and going to the "opposite" side of v's horizontal boundary edge. A *vertical stab* is defined analogously. The *stabbing information* of vertex v is the pair of boundary edges of $\mathcal{P}$ with which the horizontal and the vertical stab of v intersect first. If there is no intersection, a placeholder "phantom" line-segment

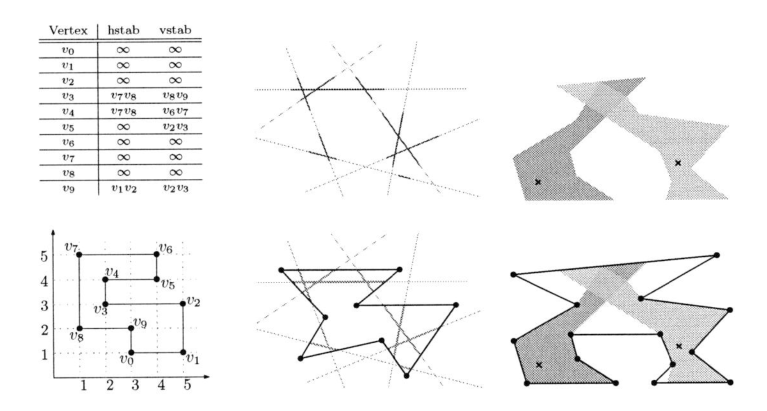

Vertex	hstab	vstab
v_0	∞	∞
v_1	∞	∞
v_2	∞	∞
v_3	v_7v_8	v_8v_9
v_4	v_7v_8	v_6v_7
v_5	∞	v_2v_3
v_6	∞	∞
v_7	∞	∞
v_8	∞	∞
v_9	v_1v_2	v_2v_3

Figure 3.1.: Constructing consistent polygons from different kinds of data. From left to right: reconstruction of orthogonal polygons from stabbing information; reconstructing of polygons from line intersections; reconstruction of polygons from visibility polygons of certain points in the interior.

is provided instead. Jackson and Wismath present an algorithm with a running time of $O(n \log n)$ that computes an orthogonal polygon $\mathcal{P}'$ of size n that is consistent with the given stabbing information at every vertex. Note that there can be more than one polygon consistent with this stabbing information.

Sidlesky et al. considered a similar reconstruction problem in which all intersections of $\mathcal{P}$ with a given set of lines $\mathcal{L}$ are known [46] (cf. Figure 3.1 (middle)). One may assume that every boundary edge of $\mathcal{P}$ is intersected by at least two lines from $\mathcal{L}$, as otherwise there are infinitely many polygons $\mathcal{P}'$ that intersect $\mathcal{L}$ in the same way that $\mathcal{P}$ does. The authors present an exponential-time algorithm that constructs all polygons $\mathcal{P}'$ that are consistent with the given intersections, including non-simple polygons.

Biedl et al. considered various types of measurements in a polygon, and considered the complexity of the problem to decide whether or not there is a polygon that is consistent with the given data [6]. Examples for the measurements considered are (1) a set of points on the boundary of the original polygon, such that every boundary edge contains at least one point, and (2) a set of visibility polygons, i.e. the regions of the polygon that are visible from certain points in the polygon (cf. Figure 3.1 (right)). The problem was shown to be NP-hard for each type of data considered in the study, if no restriction on the shape of the polygon is enforced. Polynomial-time algorithms were given for special cases only, such as when the polygon is required to be orthogonal and monotone, or star-shaped. Other special cases remain NP-hard, such as when the polygon is required to be orthogonal but not monotone.

Rappaport considered the *orthogonal-connect-the-dots* problem [44]: Given a set of points in the plane, find an orthogonal polygon whose vertices coincide with these points (cf. Figure 3.2 (left)). Note that here the points are not given in a particular order. Rappaport was able to show that the problem is NP-complete.

3.1.2. Reconstructing the Polygon Uniquely

The second variant of the reconstruction problem asks to reconstruct *the original* polygon $\mathcal{P}$ from data measured in $\mathcal{P}$. Solving this problem not only involves constructing a consistent polygon, but also requires to show that among all polygons, $\mathcal{P}$ itself is the only polygon that is consistent with the data measured in $\mathcal{P}$. This type of reconstruction problem arises naturally in the context of autonomous agents mapping polygonal environments: It is not enough for an agent to construct *some* polygon which is consistent with its observations, the agent wants to find the exact polygon it is located in. Here the focus is on the question whether reconstruction is at all possible from given data, finding efficient algorithms is only of secondary interest.

Recall the orthogonal-connect-the-dots problem mentioned above (cf. Figure 3.2 (left)). O'Rourke considered the problem further and showed that if no three consecutive vertices on the boundary of the polygon lie on a vertical or horizontal line, then the coordinates of the vertices determine the polygon uniquely [41]. What is more, he showed that in this case the problem is not NP-hard anymore, by providing an algorithm that finds the unique solution of the reconstruction problem in time $O(n \log n)$.

Coullard and Lubiw studied the problem of deciding whether a given edge-weighted graph is the *distance visibility graph* of a polygon, i.e. a visibility graph with edge-weights equal to the length of the corresponding line segments in the plane [15]. Note that this question is easy to decide in exponential time, by trying out all Hamiltonian cycles and repeatedly triangulating the graph based on the current cycle. If one of these triangulations can be embedded in the plane as a polygon, and only then, the given graph is a distance visibility graph. Furthermore, the resulting polygon is uniquely defined by the distances. Coullard and Lubiw gave a necessary condition for a graph to be the visibility graph of a polygon, and, based on this property, the authors proposed a polynomial-time algorithm that decides whether a given edge-weighted graph is the distance visibility graph of a polygon.

Snoeyink showed that every polygon $\mathcal{P}$ on n vertices is uniquely determined by its triangulation given as a graph, its *inner angles*, and its $(n-3)$ *cross-ratios* [47] (cf. Figure 3.2 (right)). A *cross-ratio* is defined in terms of quadrilaterals, i.e. pairs of triangles with a common edge in the triangulation. If a, b, c and a, d, e are the edges of the triangles of a quadrilateral in counter-clockwise order, its *cross-ratio* is the product of the lengths of b and d divided by the product of the lengths of c and e. An *inner angle* of a polygon at vertex v is the angle inside $\mathcal{P}$ which is enclosed by the two boundary edges adjacent to v (cf. Section 2.4.1).

In Chapter 5 we focus on reconstructing a polygon $\mathcal{P}$ from an ordered list of angle measurements, and we show that $\mathcal{P}$ is indeed the unique polygon consistent with this data. Moreover, we develop a polynomial-time algorithm that finds the original polygon $\mathcal{P}$.

3.2. Exploration of Graphs

In this thesis we consider agents exploring a polygon by moving along the edges of the visibility graph. The more general problem of exploring graph-like environments has received considerable attention in the past. In the following, we mention some key results in this area. Research has mostly focused on determining under which circumstances an agent can explore and reconstruct

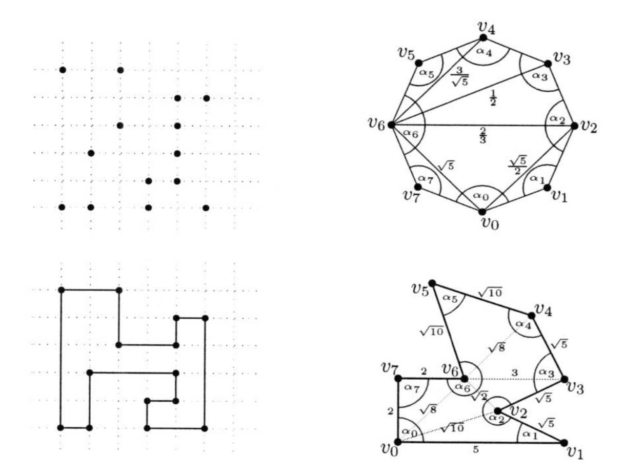

Figure 3.2.: Reconstructing a polygon uniquely from different kinds of data. Left: reconstruction of an orthogonal polygon from the coordinates of its vertices. Right: reconstruction of a polygon from the triangulation of its visibility graph with cross-ratios and inner angles. The top figure shows the triangulation of the visibility graph, where each edge is labeled by the cross-ratio of the corresponding quadrilateral. The bottom picture shows the underlying triangulation of the polygon, where each edge is labeled by its length.

an arbitrary connected graph. A very basic form of graph exploration is to traverse all the edges at least once; the corresponding problem is called the *graph traversal problem.* A stronger form of exploration is the reconstruction of (an isomorphic copy of) the graph, called the *map construction problem.* Similar to the reconstruction problem for visibility graphs, an important goal is to find minimum capabilities that the agent needs in order to solve one or both of these problems.

Intuitively, exploring a general graph-like environment is more difficult for the agent, since it cannot exploit the structure of the underlying geometry. For example, a visibility graph always contains a Hamiltonian cycle (the boundary), and we even assume the agent to be able to distinguish the edges of this cycle from other edges by their labels (recall that the boundary edges correspond to the first and last arc in the local order at each vertex). In the general setting, there is no such assumption on the labeling, except that the edges incident to each vertex are assumed to be mutually distinguishable, i.e., the graph is assumed to be locally oriented. On the other hand, in the general setting, the agent is usually assumed to be able to look back. We will see shortly that in contrast to the exploration of visibility graphs, in general graphs the ability to look back alone does not empower the agent to reconstruct the graph. Because of the ability to look back, we can model the general setting as the exploration of an undirected, edge-labeled graph of n vertices and m edges.

The exploration becomes drastically less involved when the nodes of the graph are labeled by distinct identifiers. In this case, the map construction problem is equivalent to the traversal problem, since the map can easily be constructed as soon as the identifiers of the endpoints of every edge are known. As the agent is capable of looking back, it can retrace its movements and solve the traversal problem by employing a conventional depth-first search algorithm. In fact, the depth-first traversal is asymptotically optimal in terms of the number of required moves, as it needs $\Theta(m)$ moves – obviously no other solution can do better as it has to travel along each edge at least once. A modified version of the algorithm which takes $m + O(n)$ moves has been proposed by Panaite and Pelc [43]. Another variation of the traversal problem, the so-called *piece-meal exploration*, has been studied by

Awerbuch et al. [4]. Here the agent can only execute a certain number of moves before it has to return to its home-base (e.g., for refueling).

The map construction problem becomes more challenging in an anonymous graph in which nodes are unlabeled. In this case, traversing the graph is not equivalent to map construction except for some special classes of graphs – like trees. If an upper bound on the diameter of the graph (for example n or the diameter itself) is known, traversal can always be achieved by trying all possible walks up to a length equal to this bound. Note that the knowledge of an upper bound on the diameter is necessary for terminating the traversal, without this bound the agent can still travel all edges within a finite time, but it does not know when this point is reached and can thus not stop after any finite number of steps. While we generally do not assume a bound on the memory of the agent, it was shown that it needs at least $\Omega(\log n)$ bits to traverse a graph of size n [26]. The long standing open question about the space complexity of graph traversal was closed by Reingold [45], who showed a matching upper bound of $O(\log n)$ bits on the required memory.

Even though all (connected) graphs can systematically be traversed by an agent that knows an upper bound on the number of vertices, the map construction problem cannot be solved in general. We have seen examples for indistinguishable, non-isomorphic, directed graphs in Section 2.4 (cf. Figure 2.4). We will show later that in polygonal environments, the capability of "looking back" empowers the agent to distinguish any pair of visibility graphs. Intuitively, this does not carry over to general graphs due to symmetries that can occur. As an example for indistinguishable undirected graphs, consider the pair of graphs G, H shown in Figure 3.3. Both graphs are non-isomorphic, but an agent traversing graph G makes the exact same observations as an agent traversing graph H, provided that both agents start on corresponding vertices. Thus, both these graphs are non-recognizable, i.e., the map construction problem cannot be solved for either of them. The class of all recognizable graphs has been characterized by Yamashita and Kameda [53].

One way of breaking symmetries in a graph is to equip the agent with some means of marking a node, i.e., making a node locally

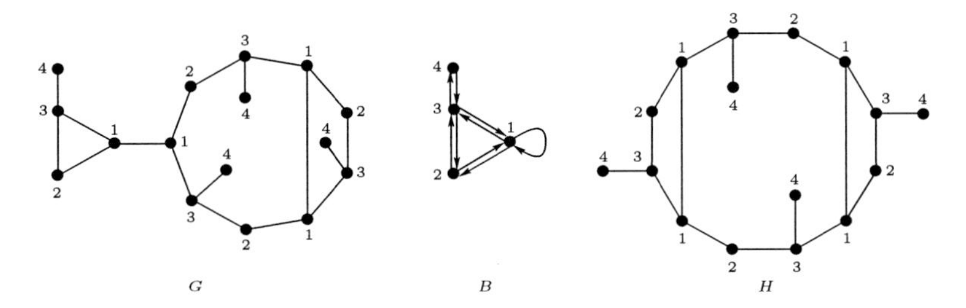

Figure 3.3.: Two non-isomorphic graphs G, H that are indistinguishable together with their minimum base graph B. The numbers on the nodes represent the equivalence classes with respect to symmetry. Edge labels are omitted for clarity.

distinguishable from all other nodes (cf. Figure 2.5). For instance, the agent may have a pebble – a simple device that can be placed on a node such that the agent recognizes the node whenever it comes back to it. A stronger model assumes that there is a *whiteboard* at each node, which allows the agent to leave information, that it can access and modify on subsequent visits of the node. The simple model with a single pebble is already enough to fully break all symmetries and enable the agent to construct a map of any graph: Starting with a map containing just the initial node, the agent extends its map one edge at a time by traversing an edge, marking the other end with the pebble, and then backtracking and checking whether the now marked node was visited before and thus already is part of the map. The model fails if there are multiple identical and indistinguishable agents (with indistinguishable pebbles) working in parallel. In this case not even a whiteboard at each node is sufficient [16].

If the agents cannot look back, i.e., in directed graphs, an agent can still always systematically traverse the graph assuming an upper bound on its size is known, by trying all possible graphs and starting locations and traversing all walks for each choice. There are families of graphs in which traversal algorithms can in fact not do better asymptotically, i.e., any algorithm requires an exponential number of moves in at least some of the graphs. Bender et al. showed that if the size of the graph is known a priori, an agent with a pebble can always construct a map of a directed, strongly connected and locally oriented graph using a polynomial number of moves [5]. Without prior knowledge on the total number of vertices n, they showed that $\Theta(\log \log n)$ pebbles are necessary and sufficient to solve the map construction problem.

We will see in Chapter 7 that an agent exploring any directed graph G can always construct the minimum base of G, provided that at least an upper bound on the number of vertices is known. When operating in an undirected graph, the same result of course carries over.

The initial knowledge of an agent does not necessarily have to be about the number of vertices n. An important question is how much prior information is necessary for mapping any general (undirected) graph. It has recently been shown that, in the worst case, solving the map construction problem can require initial knowledge of $\Omega(m \log n)$ bits [17]. This is as much information as it takes to store the graph itself, if we store a list of edges for instance.

Part I.

Boundary Exploration

The first part of the results presented in this thesis concerns agents with restricted movement capabilities. More precisely, we limit the movement of agents to be along boundary edges of the polygon only. We will refer to such restricted agents as *agents with boundary movement.* Moving along the boundary implicitly allows agents to look back, i.e., to identify where they came from after each move. Another reason for considering agents with boundary movement is that moving along the boundary allows an agent to systematically collect the observations made at every vertex. Of course, we have to counterbalance a movement restriction with extensions to the sensory of the agent if we hope to obtain agents that are able to solve the visibility graph reconstruction problem.

Definition. Let $\mathscr{F}$ be a family of arc-labeled visibility graphs. We say an agent *with boundary movement* can solve the reconstruction problem in $\mathscr{F}$ if there is an exploration strategy $\mathscr{A}$ that involves only boundary moves and that, when executed by an agent exploring any graph $G \in \mathscr{F}$, computes a graph isomorphic to G .

boundary movement

The family of visibility graphs is always implicitly fixed by the agent's capabilities, as explained in Section 2.4.1.

There is a particularly clean way of dealing with agents with boundary movement and knowledge of n, irrespective of the agent's other capabilities. In this setting, we can define a *measurement* at a vertex to be the information observed by the agent's sensors when located at the vertex. Because of the restriction to boundary moves, the data available to the agent is encoded in a sequence of measurements, one for each vertex in boundary order. Because the agent knows n, it can simply collect this data in one tour of the boundary. Any exploration strategy solving the reconstruction problem can hence be split in a *data collection phase* and a *computation phase.* The data collection phase is always the same: the agent is sent once around the boundary, collecting its observations in the process. The computation phase then only depends on the sequence of measurements and does not involve any further exploration of the polygon. Solving the reconstruction problem essentially reduces to finding an algorithm for the computation phase.

Chapter 4.

The Weakness of Combinatorial Visibilities*

In this chapter we consider agents with cvv sensor and knowledge of the total number of vertices n. We saw in Section 2.4.1 how to adapt the arc labeling of the visibility graph to account for these extensions. It will turn out that the agent model is weak in the sense that an agent that only ever moves along the boundary cannot solve the reconstruction problem. Because n is known beforehand, the agent can readily collect all the data it can hope for. This data consist of a sequence of the observations made with the cvv sensor on a tour along the boundary. The following definition makes this formal.

Definition 4.1. The *combinatorial visibility sequence* cvs of a polygon consists of the combinatorial visibility vectors of its vertices in boundary order, i.e.,

$$\mathrm{cvs} := \left(\mathrm{cvv}(v_0), \ldots, \mathrm{cvv}(v_{n-1})\right).$$

4.1. Agents with cvv Sensor

We obtain the following negative result for agents with cvv sensor.

Theorem 4.2. *An agent with boundary movement, cvv sensor, and knowledge of n cannot solve the visibility graph reconstruction problem.*

*The results presented in this chapter appeared in [7, 18].

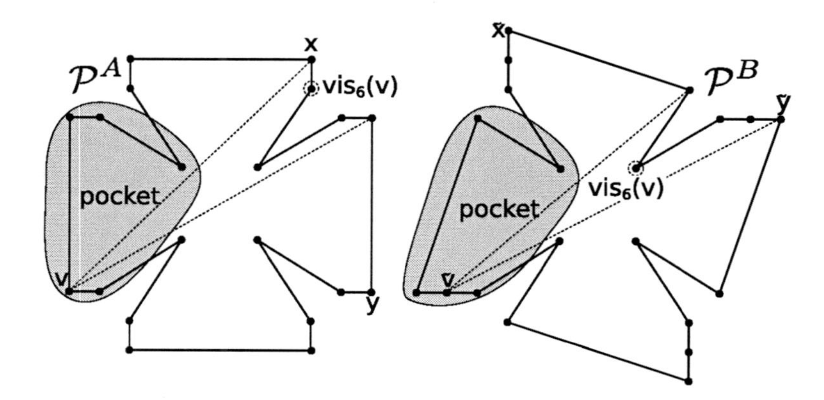

Figure 4.1.: Two polygons $\mathcal{P}^A$ and $\mathcal{P}^B$ with identical cvs and different visibility graphs.

Proof. It is sufficient to provide two polygons with different visibility graphs, but with the same numbers of vertices and with the same combinatorial visibility sequence. In Figure 4.1 we present two such polygons $\mathcal{P}^A$ and $\mathcal{P}^B$. The proof is by inspection of the polygons together with the list of cvv's and sequences of visible vertices for each vertex in Figure 4.2. Note that our construction is not in general position in the sense that it contains collinear triples of points, however the polygons can easily be perturbed slightly without changing visibilities or cvv's.

The idea behind the construction of the polygons is to use multiple copies of a "pocket" of vertices (cf. Figure 4.1 for an illustration). Each pocket forms a convex curve, but the vertices connecting the pockets form reflex angles, resulting in a non-convex polygon $\mathcal{P}$. The vertices inside a pocket thus do not see all vertices of $\mathcal{P}$, they see (apart from their own pocket) only parts of exactly two pockets. We use the fact that the vertices have no way to distinguish what pockets they are "looking into" and we modify the polygon $\mathcal{P}^A$ by shifting the vertex c (cf. Figure 4.2) so that in $\mathcal{P}^B$ the shifted vertex $\tilde{c}$ looks into different pockets, while not changing the combinatorial visibility vector of any vertex.

We determined experimentally, by exhaustive enumeration of visibility graphs, that no counterexample exists with ten or fewer

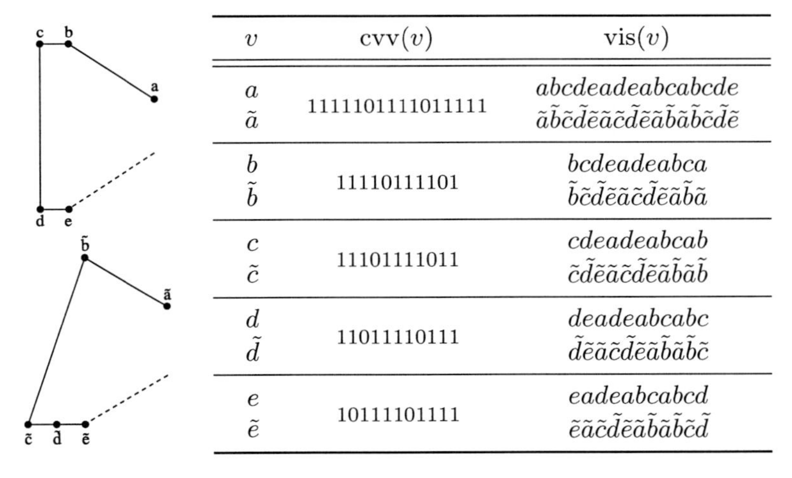

v	$\text{cvv}(v)$	$\text{vis}(v)$
a	1111101111011111	$abcdeadeabcabcde$
$\tilde{a}$		$\tilde{a}\tilde{b}\tilde{c}\tilde{d}\tilde{e}\tilde{a}\tilde{c}\tilde{d}\tilde{e}\tilde{a}\tilde{b}\tilde{a}\tilde{b}\tilde{c}\tilde{d}\tilde{e}$
b	11110111101	$bcdeadeabca$
$\tilde{b}$		$\tilde{b}\tilde{c}\tilde{d}\tilde{e}\tilde{a}\tilde{c}\tilde{d}\tilde{e}\tilde{a}\tilde{b}\tilde{a}$
c	11101111011	$cdeadeabcab$
$\tilde{c}$		$\tilde{c}\tilde{d}\tilde{e}\tilde{a}\tilde{c}\tilde{d}\tilde{e}\tilde{a}\tilde{b}\tilde{a}\tilde{b}$
d	11011110111	$deadeabcabc$
$\tilde{d}$		$\tilde{d}\tilde{e}\tilde{a}\tilde{c}\tilde{d}\tilde{e}\tilde{a}\tilde{b}\tilde{a}\tilde{b}\tilde{c}$
e	10111101111	$eadeabcabcd$
$\tilde{e}$		$\tilde{e}\tilde{a}\tilde{c}\tilde{d}\tilde{e}\tilde{a}\tilde{b}\tilde{a}\tilde{b}\tilde{c}\tilde{d}$

Figure 4.2.: The sequences $\text{cvv}(c)$ and $\text{vis}(v)$ of every vertex within a pocket of $\mathcal{P}^A$ and $\mathcal{P}^B$, where a, b, c, d, e each refer to all four vertices at the corresponding position within their pocket in $\mathcal{P}_A$ and $\tilde{a}, \tilde{b}, \tilde{c}, \tilde{d}, \tilde{e}$ refer to their counterparts in $\mathcal{P}_B$.

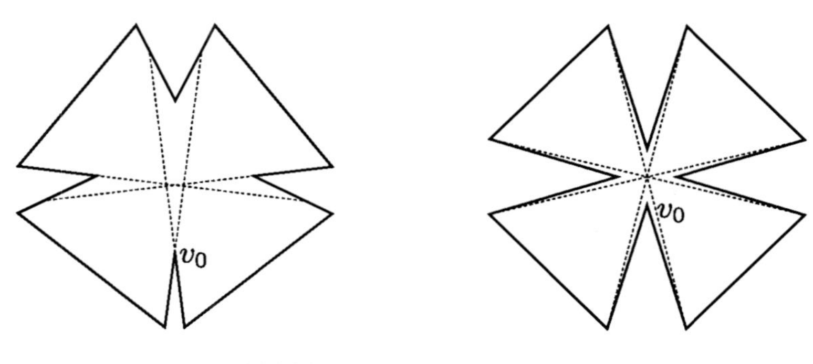

$$\mathbf{cvs} = 1110110111, 110101, 101011, \ldots$$

Figure 4.3.: Two polygons with twelve vertices each that have identical combinatorial visibility sequences and different visibility graphs.

vertices. Figure 4.3 gives another example with twelve vertices. The example essentially uses the same principle as the one above, but its construction is somewhat more tedious. □

Theorem 4.2 shows that the knowledge of the combinatorial visibility sequence is not sufficient to reconstruct the visibility graph of a polygon. A natural question is how to extend this information "minimally" in order to make the reconstruction possible. In the following we show that adding the knowledge of the interior angles of a polygon is still not enough. We prove the following theorem.

Theorem 4.3. *An agent with boundary movement, cvv sensor, inner-angle sensor, and knowledge of n cannot solve the visibility graph reconstruction problem.*

Proof. Figure 4.4 shows a modified version of the polygons $\mathcal{P}^A$ and $\mathcal{P}^B$ of Figure 4.1. As one can easily check, the polygons still have the same combinatorial visibility sequence and different visibility graphs. In addition, they also have the same inner angles at the vertices. The existence of such polygons proves the theorem. □

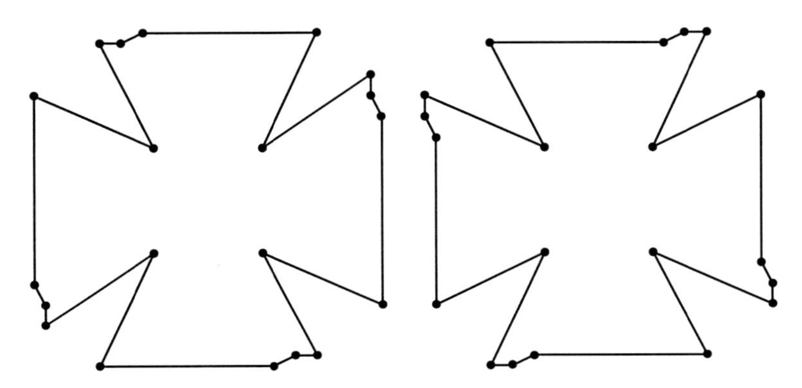

Figure 4.4.: Two polygons with identical combinatorial visibility sequences and identical interior angles but different visibility graphs. The visibilities are similar to those of $\mathcal{P}^A$ and $\mathcal{P}^B$.

Note that Theorems 4.2 and 4.3 do not imply that an agent that is not constrained to move along the boundary and is equipped with sensors for measuring cvv's and/or inner angles cannot reconstruct the visibility graph. Such an agent would be able to distinguish $\mathcal{P}^A$ and $\mathcal{P}^B$ by moving to a specific vertex of a distant pocket and inspecting its cvv. However it seems difficult for such a agent to reconstruct the visibility graph in general – the question whether this is always possible remains open.

4.2. Periodical Visibility Sequence

The following theorem considers a related question about polygons with periodical combinatorial visibility sequence S. Let v_i and v_j be two vertices corresponding to two periodical partners in S. The question is whether the l-th vertex visible to v_i and the l-th vertex visible to v_j need to correspond to periodical partners in S themselves. The existence of such a property would have an impact on various interesting problems in the field of polygon exploration; for example, on the weak rendezvous problem for two agents in symmetrical polygons. We show however that the property does not hold.

Theorem 4.4. *There is a polygon $\mathcal{P}$ with periodical combinatorial visibility sequence S of period p, a vertex v_i of $\mathcal{P}$, and an index $x \in [d(v_i)]$ for which*

$$\mathrm{cvv}(\mathrm{vis}_x(v_i)) \neq \mathrm{cvv}(\mathrm{vis}_x(v_{i+p})).$$

Proof. We construct a polygon $\mathcal{P}$ with visibility sequence of period $p = n/2$ and with the aforementioned property from the two polygons $\mathcal{P}^A$ and $\mathcal{P}^B$ in Figure 4.1. The construction can then easily be generalized to $p < n/2$.

The idea of the construction is to "glue" $\mathcal{P}^A$ and $\mathcal{P}^B$ together at vertices v and $\tilde{v}$ of $\mathcal{P}^A$ and $\mathcal{P}^B$, respectively, where v and $\tilde{v}$ are as depicted in Figure 4.1. We want to glue the polygons such that every two corresponding vertices w and $\tilde{w}$ of the two polygons are periodical partners in S. Thus, we need to glue the polygons such that the cvv's of corresponding vertices w and $\tilde{w}$ are the same. We can then use the result of Theorem 4.2 which guarantees the existence of vertices w and $\tilde{w}$ with the same cvv but different vision. Formally, if w from $\mathcal{P}^A$ is a vertex v_i in $\mathcal{P}$ and $\tilde{w}$ from $\mathcal{P}^B$ is a vertex $v_{i+n/2}$ in $\mathcal{P}$, there is an index x such that $\mathrm{vis}_x(v_i) = v_k$ and $\mathrm{vis}_x\left(v_{i+\frac{n}{2}}\right) = v_l \neq v_{k+\frac{n}{2}}$. Because of the structure of the two polygons, we will have $\mathrm{cvv}(v_k) \neq \mathrm{cvv}(v_l)$ which proves the theorem.

The problem when gluing at v and $\tilde{v}$ is that these vertices have to be split in the process, which makes them appear different than all other vertices. By inserting spikes (cf. Figure 4.5) at all vertices, we can again make vertices appear the same from afar while still maintaining equal combinatorial visibility sequences. Spikes can easily be inserted at convex vertices such that no distant vertex is visible from the spike tip and the spike tip's neighbors retain the vision of the original vertex (except for seeing the former vertices as gaps and seeing the new spike tip). It is however not generally clear how to do the same for reflex vertices, Figure 4.5 shows how it can be accomplished for the four reflex vertices in our case. Figure 4.6 shows the spiked versions of $\mathcal{P}^A$ and $\mathcal{P}^B$ before gluing. Figure 4.7 lists how the cvv's change with the introduction of spikes.

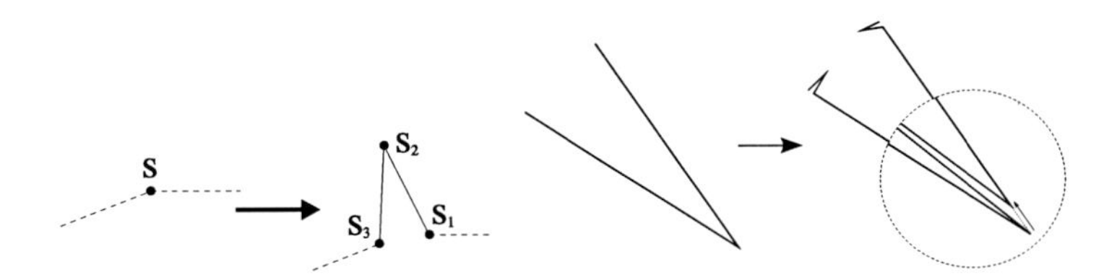

Figure 4.5.: Left: The concept of inserting spikes at vertices. Right: Illustration of how the spikes are inserted at reflex vertices. We chose our modification such that the right neighbor of the spike tip retains the visibility of the original vertex.

Once we have spiked versions of $\mathcal{P}^A$ and $\mathcal{P}^B$, we can glue them together in a straightforward way by simply splitting the spike tip of v and $\tilde{v}$ and attaching the open ends to one another. It can easily be seen that the gluing produces a periodical visibility sequence S. Figure 4.8 shows the resulting polygon $\mathcal{P}$. The extension to $p < n/2$ is easily made, as we can attach more than two copies of the two spiked polygons around a common center (cf. Figure 4.9). Note that we could also have used the smaller polygons from Figure 4.3 for gluing, but then the process becomes more tedious (cf. Figure 4.10). □

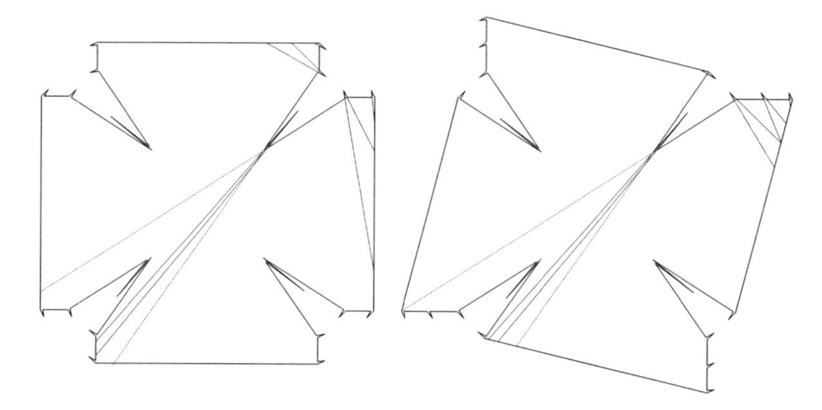

Figure 4.6.: The two polygons from Figure 4.1 equipped with spikes and still with identical combinatorial visibility sequences. The areas visible from the different spike-tips are indicated.

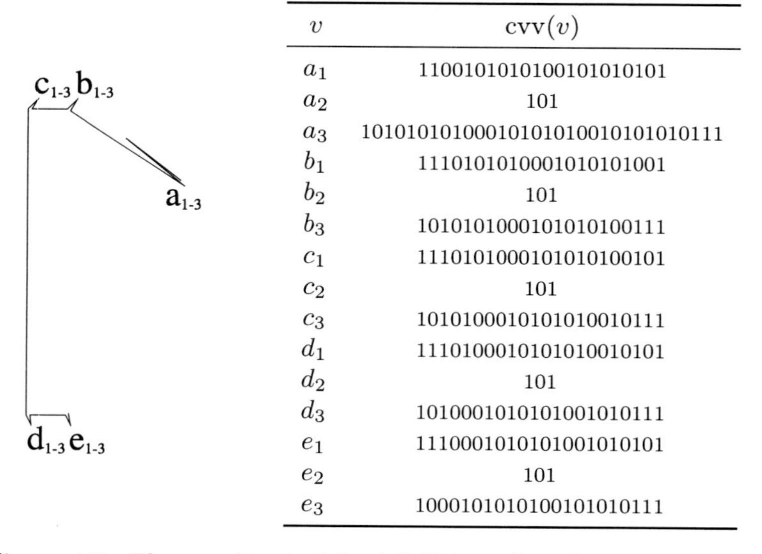

v	cvv(v)
a_1	110010101010010101010101
a_2	101
a_3	1010101010001010101010010101010111
b_1	111010101000101010101001
b_2	101
b_3	101010100010101010100111
c_1	111010100010101010100101
c_2	101
c_3	101010001010101010010111
d_1	111010001010101010010101
d_2	101
d_3	101000101010101001010111
e_1	111000101010101001010101
e_2	101
e_3	100010101010100101010111

Figure 4.7.: The combinatorial visibilities of each vertex in a pocket of $\mathcal{P}^A$ after adding spikes (the same cvv's arise for $\mathcal{P}^B$). We write v_{1-3} to denote the group of vertices v_1, v_2, v_3.

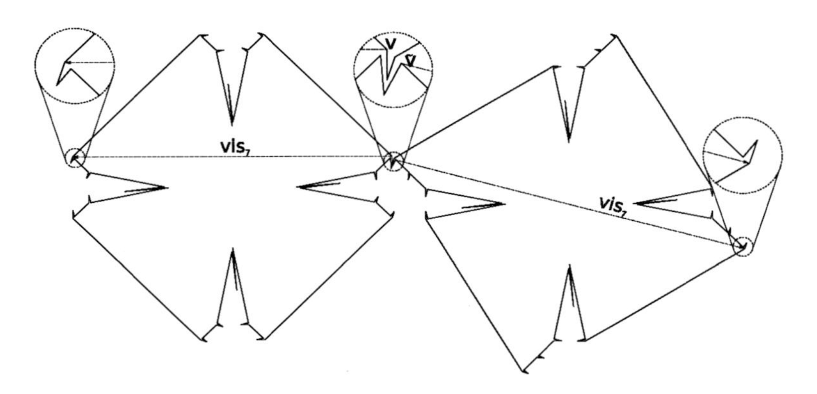

Figure 4.8.: The polygon with $n = 120$ that proves Theorem 4.4.

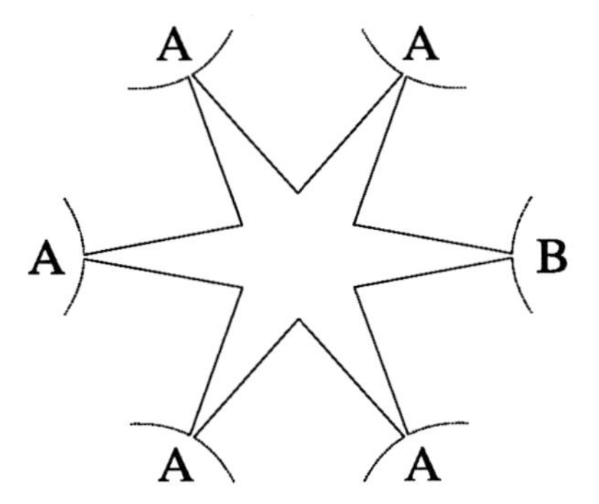

Figure 4.9.: Illustration of how to glue multiple copies of two spiked polygons in order to obtain a polygon with periodical combinatorial visibility sequence.

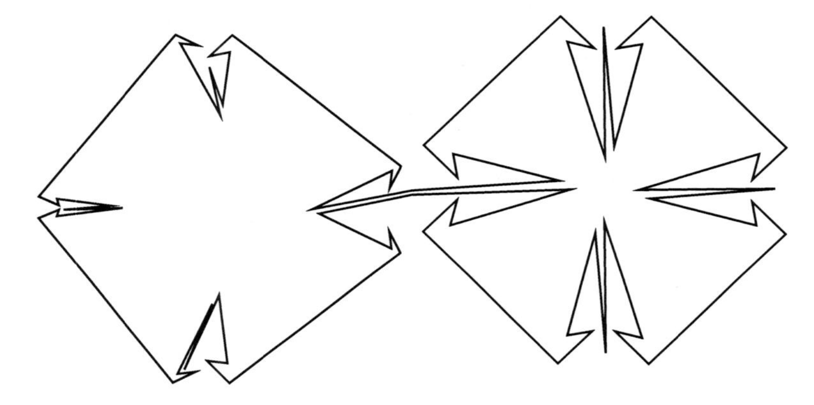

Figure 4.10.: An alternative example for the proof of Theorem 4.4 with $n = 72$.

Chapter 5.

Mapping with Angles

Intuitively, in order to obtain an agent model that allows to solve the visibility graph reconstruction problem, we need to counterbalance the movement restriction to boundary movements by giving the agents powerful sensing capabilities. In the previous chapter we saw that the ability to measure interior angles is insufficient, even when combined with the cvv sensor and knowledge of n. In this chapter we consider agents equipped with an angle sensor. We show that such agents can solve the visibility graph reconstruction problem. We start by proving this for the case that n is known to the agent beforehand. Afterwards we will show how the agent can accomplish the reconstruction without this additional knowledge.

5.1. Reconstruction when Knowing n*

As explained before, knowledge of n makes it easy for the agent to separate the reconstruction problem into a collection and a reconstruction phase. The data that the agent obtains during a tour of the boundary can be described as a sequence of *angle measurements.* An angle measurement for a vertex v is defined as the sequence $(\angle_v(1,2), \angle_v(2,3), \ldots, \angle_v(d(v)-1, d(v)))$. Obviously, an angle measurement implicitly encodes the angle between any two arcs at v (cf. Figure 5.1). For the remainder of this section, we can forget about the agent and focus on the problem whether a sequence of angle measurements contains enough information to uniquely infer the visibility graph from which it originates. Note

*The results presented in this section appeared in [19, 21].

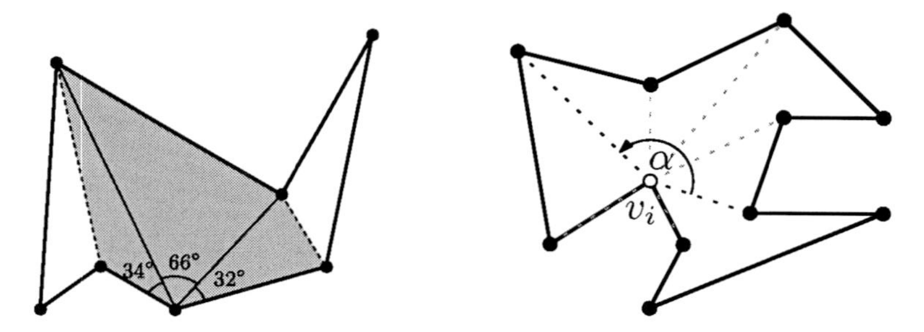

Figure 5.1.: Left: An angle measurement at a vertex yields the list of angles between adjacent edges of G_{vis} in boundary order, here $(32°, 66°, 34°)$. Right: The angle between non-adjacent edges can easily be computed by summing the angles in between.

that we assume angle measurements to be ordered according to the perception of the agent. Figure 5.2 shows that this order is necessary in order to reconstruct the visibility graph.

Once we have shown that the visibility graph G_{vis} can uniquely be reconstructed from the sequence of angle measurements, it follows that we can also uniquely reconstruct the shape of the corresponding polygon. This is because from G_{vis} we can find the set of edges of the triangles in some triangulation of the polygon. From the angle data we can then infer the shape of each triangle up to scaling, and hence the shape of the polygon. Note that we can do this in linear time, as we can find a triangulation in linear time using Proposition 2.19. It remains to argue that independent of the triangulation, the resulting polygon is the same. This is due to the fact that every possible triangulation uniquely leads to a polygon, and this polygon in turn admits every single one of the possible triangulations. Of course, conversely, the visibility graph can be computed from the shape of the polygon. The visibility graph reconstruction problem therefore is equivalent to the reconstruction of the polygon shape (cf. Figure 5.3).

The key question when trying to reconstruct the visibility graph of a polygon is how to identify a vertex visible to a known vertex v. Knowing all angles at every vertex may seem to be more information than necessary and the reconstruction problem may

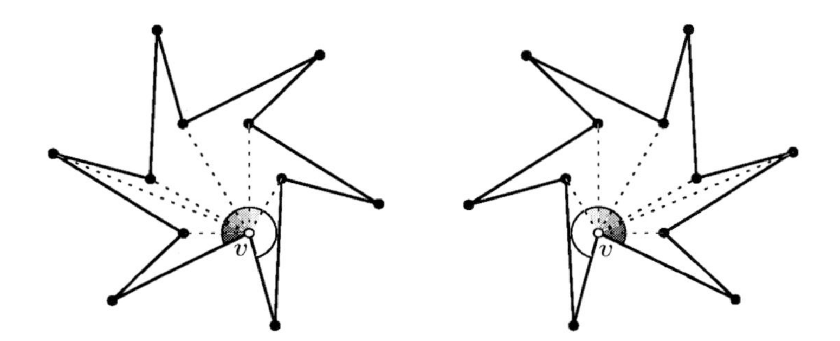

Figure 5.2.: Two polygons with different visibility graphs and the same set of angles at every vertex.

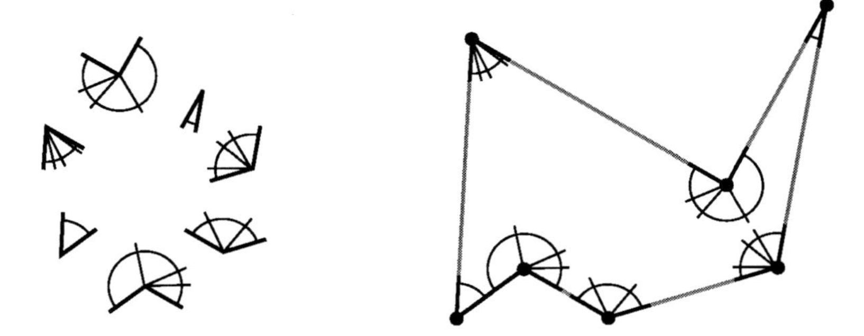

Figure 5.3.: Given the angle measurement for each vertex in counter-clockwise order (left), the goal is to find the unique polygon shape that fits these angle measurements (right).

thus seem easily solvable by some greedy algorithm. Before we actually present an algorithm that solves the reconstruction problem, we show that some natural greedy algorithms do not work in general.

5.1.1. A Greedy Approach

It is a natural idea to first orient all angles with respect to a single, global orientation (e.g. the line $\overline{v_{n-1}v_0}$) by summing angles around the polygon boundary. Then, if a vertex v sees some other vertex u under a certain global angle α, u must see v under the inverse angle $\alpha + \pi$, as the line $\overline{uv}$ has a defined orientation. A simple greedy approach to identify the vertex u in the view from v would be to inspect vertices along the boundary starting from v and find the first vertex in boundary order that sees some other vertex under the global angle $\alpha + \pi$. The example in Figure 5.4 shows that this approach does not work in general.

A similar but somewhat more rigorous approach is to allow global angles to go beyond $[0, 2\pi)$ while summing around the polygon boundary (cf. Figure 5.4). This prevents pairing vertex v_0 with vertex v_1 in the example. Nevertheless, there are still examples where this strategy fails, and in fact it is not possible at all to greedily match angles: Inspect Figure 5.5 for an example of two polygons for which any greedy way of pairing up the vertices vertices has to fail for one of the two.

5.1.2. Triangle Witness Algorithm

We now give an algorithm for the reconstruction of a visibility graph $G_{\text{vis}} = (V, A)$ from the angles at each vertex. Note that from now on we map all angles to the range $[0, 2\pi)$. Our algorithm considers all vertices at once and incrementally identifies edges connecting vertices that lie further and further apart along the boundary. In step k of the algorithm, for each vertex v_i, we know which vertices in $\{v_{i+1}, v_{i+2}, \dots, v_{k-1}\}$ are visible to v_i and need to decide whether or not v_i sees v_k. Intuitively, the decision boils down to the question whether the next unidentified vertex in vis(v_i) is v_k. Our algorithm only needs to make decisions of this type. The key ingredient here is the use of a *triangle witness*

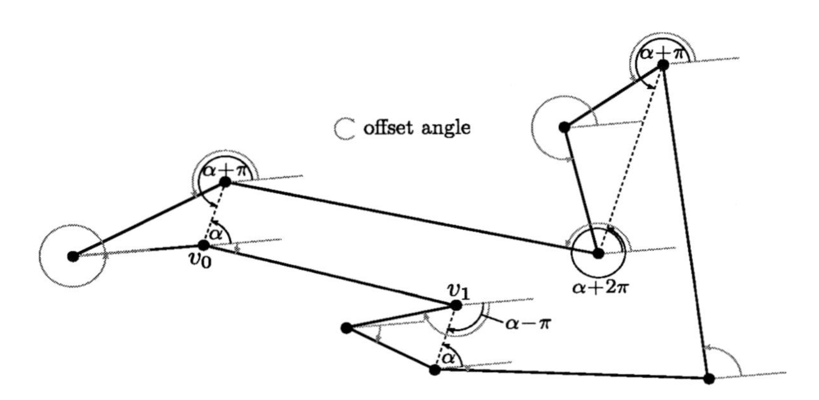

Figure 5.4.: Illustration of the idea behind the greedy pairing algorithm for a single angle α and starting vertex v_0. If we map angles to the range $[0, 2\pi)$, we allow v_0 and v_1 to be paired which is obviously a mistake.

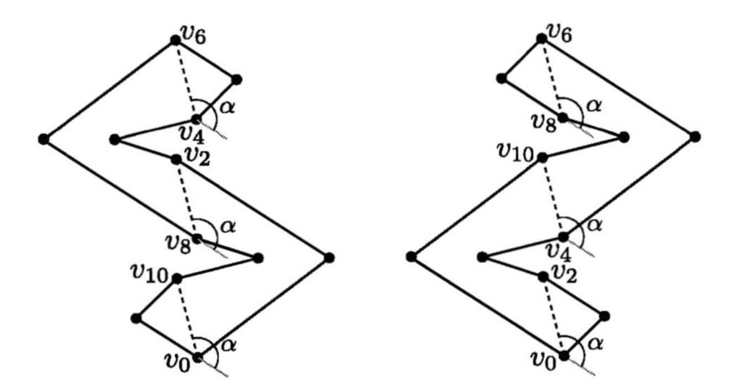

Figure 5.5.: An example in which only one visibility graph can correctly be reconstructed by any greedy pairing algorithm.

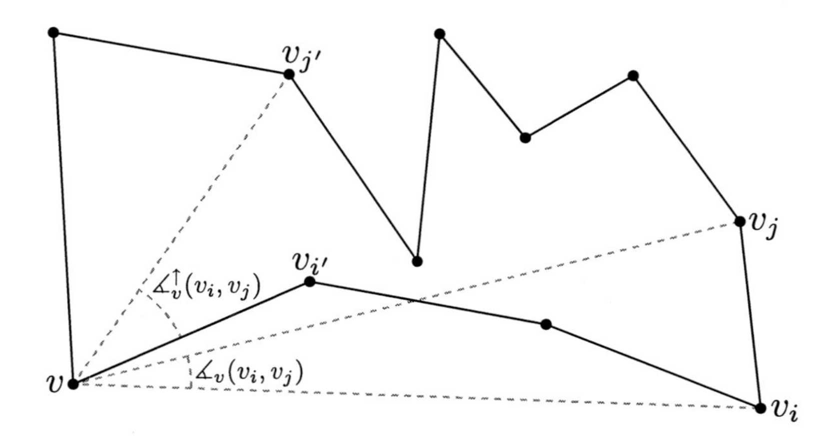

Figure 5.6.: Illustration of the approximation $\measuredangle_v^{\uparrow}(v_i, v_j) = \measuredangle_v(v_{i'}, v_{j'})$ of the angle $\measuredangle_v(v_i, v_j)$.

vertex that indicates whether two other vertices see each other. Because any polygon can be triangulated, we know that for every two vertices v_i, v_j, with $v_j \neq v_{i+1}$ and $(v_i, v_j) \in A$, there is a "witness" vertex $v_l \in \text{chain}(v_{i+1}, v_{j-1})$ that they both see, such that v_i, v_l, and v_j form a triangle with an angle sum of π. We now extend this notion to the case where $(v_i, v_j) \notin A$. But first we will need the approximation $\measuredangle_v^{\uparrow}(v_i, v_j)$ of the angle $\measuredangle_v(v_i, v_j)$ for vertex v, which is defined as follows (cf. Figure 5.6). We use $\text{chain}_v(v_i, v_j)$ to denote the longest subsequence of $\text{chain}(v_i, v_j)$ that contains only vertices visible to v.

$\measuredangle_v^{\uparrow}(v_i, v_j)$

Definition 5.1. For any $v, v_i, v_j \in V$, let $v_{i'}$ be the last vertex in $\text{chain}_v(v_{i+1}, v_i)$ and $v_{j'}$ be the first vertex in $\text{chain}_v(v_j, v_{j-1})$. We define $\measuredangle_v^{\uparrow}(v_i, v_j) := \measuredangle_v(v_{i'}, v_{j'})$.

Observe that if $(v, v_i), (v, v_j) \in A$, we have $\measuredangle_v^{\uparrow}(v_i, v_j) = \measuredangle_v(v_i, v_j)$. With this in mind, we can define a generalized condition.

triangle witness

Definition 5.2. Let $v_i, v_j \in V$ be two different vertices and $v_j \neq v_{i+1}$. Let further $v_l \in \text{chain}(v_{i+1}, v_{j-1})$ with $(v_i, v_l), (v_j, v_l) \in A$. We say v_l is a *triangle witness* of (v_i, v_j) if it fulfills the *generalized angle-sum condition* (cf. Figure 5.7)

$$\measuredangle_{v_i}^{\uparrow}(v_l, v_j) + \measuredangle_{v_j}^{\uparrow}(v_i, v_l) + \measuredangle_{v_l}(v_j, v_i) = \pi.$$

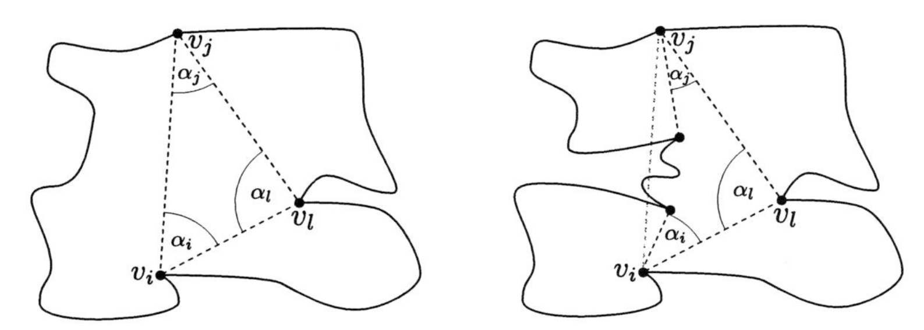

Figure 5.7.: Illustration of the generalized angle sum condition of Definition 5.2. On the left $(v_i, v_j) \in A$ and the three angles $\alpha_i = \measuredangle^{\uparrow}_{v_i}(v_l, v_j)$, $\alpha_j = \measuredangle^{\uparrow}_{v_j}(v_i, v_l)$, $\alpha_l = \measuredangle_{v_l}(v_j, v_i)$ of the condition sum up to π – hence v_l is a triangle witness of (v_i, v_j). On the right, $(v_i, v_j) \notin A$ and the sum of the angles is strictly less than π – hence v_l is no triangle witness of (v_i, v_j).

We will show later that two vertices $v_i, v_j \in V$, $|\text{chain}(v_i, v_j)| > 2$, see each other if and only if there is a triangle witness of (v_i, v_j). Note that if $v_j = v_{i+1}$, the *ordered* pair (v_i, v_j) does not have a triangle witness even though $(v_i, v_j) \in A$. Let us briefly motivate the generalized angle-sum condition. As before, we know that if two vertices $v_i, v_j \in V, v_j \neq v_{i+1}$, see each other, there must be a vertex $v_l \in \text{chain}(v_{i+1}, v_{j-1})$ which sees both of them. For any such choice of v_l, the condition $\measuredangle_{v_i}(v_l, v_j) + \measuredangle_{v_j}(v_i, v_l) + \measuredangle_{v_l}(v_j, v_i) = \pi$ is trivially fulfilled. In the case that v_i does not see v_j, the only difference from v_i's perspective is that for any choice of v_l, the angle between $\overline{v_i v_l}$ and $\overline{v_i v_j}$ does not appear in v_i's angles $\angle_{v_i}$ (although another angle with the same value might still appear in $\angle_{v_i}$). In order to capture this difference, we replace v_j in $\measuredangle_{v_i}(v_l, v_j)$ by an expression that evaluates to v_j if and only if v_i sees v_j. We choose the expression "the first vertex in $\text{chain}_{v_i}(v_j, v_{j-1})$", which is v_j exactly if v_i sees v_j. If, similarly, we also replace v_i in $\measuredangle_{v_j}(v_i, v_l)$ by "the last vertex in $\text{chain}_{v_j}(v_{i+1}, v_i)$", we obtain the generalized angle-sum condition of Definition 5.2.

We can now describe the triangle witness algorithm. It iterates over an increasing step-size k along the boundary, focusing in iterations k on all arcs of the form (v_i, v_{i+k}) and (v_{i+k}, v_i). Throughout, it maintains two maps F,B that store for every vertex all the arcs identified so far that go at most k steps forward or backward along the boundary, respectively. Both $F[v_i]\,[v_j] = s$ and $B[v_i]\,[v_j] = s$ denote that (v_i, v_j) is the s-th arc at v_i in boundary order. The difference between $F[v_i]$ and $B[v_i]$ is that $B[v_i]$ is filled in reversed boundary order by the algorithm, i.e., its first entry will be $B[v_i]\,[v_{i-1}] = d(v_i)$. Whenever convenient, we use $F[v_i]$ and $B[v_i]$ like a set, e.g., we write $v_l \in F[v_i]$ to denote that there is an entry v_l in $F[v_i]$ and write $|F[v_i]|$ to denote the number of entries in $F[v_i]$. It is clear that once we computed the maps for k up to $\lceil \frac{n}{2} \rceil$, we essentially have computed A.

The initialization of the maps for $k = 1$ is simple as every vertex sees its neighbors on the boundary. In later iterations, for every vertex v_i there is always exactly one candidate vertex for v_{i+k}, namely the $(|F[v_i]| + 1)$-th vertex visible to v_i. We decide whether or not v_i and v_{i+k} see each other, by going over all vertices between v_i and v_{i+k} in boundary order along the boundary and checking whether there is a triangle witness $v_l \in \text{chain}(v_{i+1}, v_{i+k-1})$ of (v_i, v_{i+k}). If and only if this is the case, we update A, F, and B with the arcs (v_i, v_{i+k}) and (v_{i+k}, v_i). For a listing of the triangle witness algorithm see Algorithm 1.

In the following we prove the correctness of the triangle witness algorithm. For this we mainly have to show that having a triangle witness is necessary and sufficient for a pair of vertices to see each other. We first give the following lemma.

Lemma 5.3. *Let $v_i, v_j \in V$ with $i = j + 2$ be two vertices that do not see each other. If $w := v_{j+1} = v_{i-1}$ is convex, we have that both $v_{j'} = \arg\min_{v_b \in \text{chain}_{v_i}(v_{i+1}, v_{j-1})} \measuredangle_{v_i}(v_b, w)$ as well as $v_{i'} = \arg\min_{v_b \in \text{chain}_{v_j}(v_{i+1}, v_{j-1})} \measuredangle_{v_j}(w, v_b)$ are blockers of (v_i, v_j) and lie in the interior of the triangle $\triangle_{v_i v_j w}$.*

Proof. As w is convex, the shortest path p_{ij} from v_i to v_j only contains vertices of $\text{chain}(v_i, v_j)$. As p_{ij} either only makes left or only makes right turns (depending on the boundary order), all interior vertices of p_{ij} lie in the interior of $\triangle_{v_i v_j w}$. Furthermore $v_{j'}$ and $v_{i'}$ are the first and the last interior vertices of p_{ij} respectively. By Proposition 2.37 we have that both $v_{j'}$ and $v_{i'}$

Algorithm 1: Triangle witness algorithm.

input : n, $d(\cdot)$, $\angle_{\cdot}(\cdot,\cdot)$
output: the set of arcs A

1 F,$B \leftarrow$ [array of n empty maps], $A \leftarrow \emptyset$
2 **for** $i \leftarrow 0, \ldots, n-1$ **do**
3 $\quad A \leftarrow A \cup \{(v_i, v_{i+1}), (v_{i+1}, v_i)\}$
4 $\quad F[v_i]\,[v_{i+1}] \leftarrow 1$
5 $\quad B[v_{i+1}]\,[v_i] \leftarrow d(v_i)$
6 **for** $k \leftarrow 2, \ldots, \lceil \frac{n}{2} \rceil$ **do**
7 $\quad$ **for** $i \leftarrow 0, \ldots, n-1$ **do**
8 $\quad\quad j \leftarrow i + k$
9 $\quad\quad$ **for** $l \leftarrow i+1, \ldots, j-1$ **do**
10 $\quad\quad\quad$ **if** $v_l \in F[v_i] \wedge v_l \in B[v_j]$ **then**
11 $\quad\quad\quad\quad \alpha_i \leftarrow \angle_{v_i}(F[v_i]\,[v_l]\,, |F[v_i]| + 1)$
$\quad\quad\quad\quad (= \measuredangle^{\uparrow}_{v_i}(v_l, v_j)$, cf. Theorem 5.5)
12 $\quad\quad\quad\quad \alpha_j \leftarrow \angle_{v_j}(d(v_j) - |B[v_j]|\,, B[v_j]\,[v_l])$
$\quad\quad\quad\quad (= \measuredangle^{\uparrow}_{v_j}(v_i, v_l)$, cf. Theorem 5.5)
13 $\quad\quad\quad\quad \alpha_l \leftarrow \angle_{v_l}(F[v_l]\,[v_j]\,, B[v_l]\,[v_i])$
$\quad\quad\quad\quad (= \measuredangle_{v_l}(v_j, v_i)$, cf. Theorem 5.5)
14 $\quad\quad\quad\quad$ **if** $\alpha_i + \alpha_j + \alpha_l = \pi$ **then**
15 $\quad\quad\quad\quad\quad A \leftarrow A \cup \{(v_i, v_j), (v_j, v_i)\}$
16 $\quad\quad\quad\quad\quad F[v_i]\,[v_j] = |F\,[v_i]| + 1$
17 $\quad\quad\quad\quad\quad B[v_j]\,[v_i] = d(j) - |B[v_j]|$
18 $\quad\quad\quad\quad\quad$ abort innermost loop

are blockers of (v_i, v_j) or of (v_j, v_i). But since all interior vertices from p_{ij} are in chain(v_i, v_j), it follows that $v_{j'}$ and $v_{i'}$ are blockers of (v_i, v_j). □

Now to the central lemma:

Lemma 5.4. *Let $v_i, v_j \in V$ with $|\text{chain}(v_i, v_j)| > 2$. There is a triangle witness v_l of (v_i, v_j) if and only if v_i sees v_j.*

Proof. Assume v_i sees v_j. By Proposition 2.16 there exists a triangulation of the polygon using the line segment $\overline{v_i v_j}$. Hence there must be a vertex $v_l \in \text{chain}(v_{i+1}, v_{j-1})$ for which both (v_i, v_l) and (v_l, v_j) are in A. For this vertex we have $\measuredangle^{\uparrow}_{v_i}(v_l, v_j) + \measuredangle^{\uparrow}_{v_j}(v_i, v_l) + \measuredangle_{v_l}(v_j, v_i) = \measuredangle_{v_i}(v_l, v_j) + \measuredangle_{v_j}(v_i, v_l) + \measuredangle_{v_l}(v_j, v_i) = \pi$ as all three relevant edges are lines of sight in the polygon, and because the sum over the angles of any triangle is π.

For the converse implication assume there is a triangle witness v_l of (v_i, v_j). For the sake of contradiction, assume v_i does not see v_j.

Consider the polygon $\mathcal{P}'$ induced by the closed polygonal chain $(v_i, v_l, v_j) \oplus \text{chain}(v_{j+1}, v_{i-1})$ (cf. Figure 5.8). As $(v_i, v_l), (v_l, v_j) \in A$, $\mathcal{P}'$ is simple and well defined by Proposition 2.8. In $\mathcal{P}'$, v_l is a convex vertex: By assumption, v_l fulfills the generalized angle-sum condition of Definition 5.2 and thus $\measuredangle_{v_l}(v_j, v_i) \leq \pi$, because all angles are non-negative. We can therefore apply Lemma 5.3 (on v_j, v_i) with respect to $\mathcal{P}'$ and conclude that both $v_{j'}$ and $v_{i'}$ block (v_j, v_i), where $v_{j'} = \arg\min_{v_b \in \text{chain}_{v_i}(v_{j+1}, v_{i-1})} \measuredangle_{v_i}(v_l, v_b)$ and $v_{i'} = \arg\min_{v_b \in \text{chain}_{v_j}(v_{j+1}, v_{i-1})} \measuredangle_{v_j}(v_b, v_l)$. This is then also true in our original polygon $\mathcal{P}$ and thus $v_{i'} \in \text{chain}(v_j, v_{j'})$ as otherwise $v_{j'}$ would block $(v_j, v_{i'})$ and $v_{i'}$ would block $(v_{j'}, v_i)$ contradicting the definition of $v_{j'}$ and $v_{i'}$, respectively. Observe that $v_{i'}$ is the last vertex in chain(v_{i+1}, v_i) visible to v_j and $v_{j'}$ is the first vertex in chain(v_j, v_{j-1}) visible to v_i.

By applying Lemma 5.3 to $\mathcal{P}'$, we know that both $v_{j'}$ and $v_{i'}$ lie in the interior of $\triangle_{v_i v_l v_j}$. This means $\measuredangle^{\uparrow}_{v_i}(v_l, v_j) = \measuredangle_{v_i}(v_l, v_{j'}) < \measuredangle_{v_i}(v_l, v_j)$ and $\measuredangle^{\uparrow}_{v_j}(v_i, v_l) = \measuredangle_{v_j}(v_{i'}, v_l) < \measuredangle_{v_j}(v_i, v_l)$ and thus $\measuredangle^{\uparrow}_{v_i}(v_l, v_j) + \measuredangle^{\uparrow}_{v_j}(v_i, v_l) + \measuredangle_{v_l}(v_j, v_i) < \measuredangle_{v_i}(v_l, v_j) + \measuredangle_{v_j}(v_i, v_l) + \measuredangle_{v_l}(v_j, v_i) = \pi$, which is a contradiction with our assumption that v_l is a triangle witness of (v_i, v_j). □

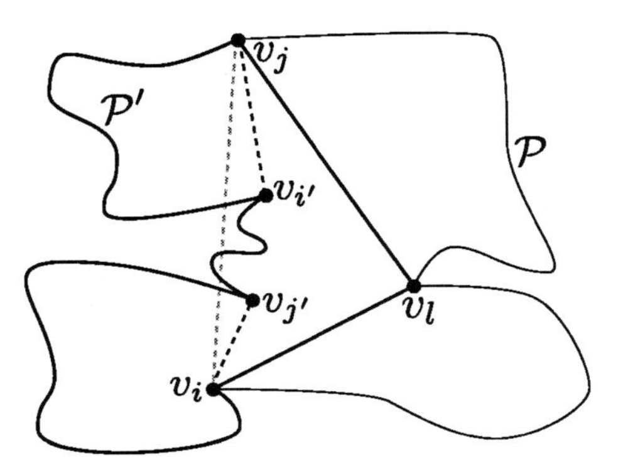

Figure 5.8.: Sketch of the definitions in the proof of Lemma 5.4.

Using this, we can now prove the main result of this section.

Theorem 5.5. *The triangle witness algorithm is correct and computes a unique solution.*

Proof. As the arcs in A are the same as the arcs stored in F and the same as the arc stored in B throughout the algorithm, it is sufficient to show that after step k of the iteration both F and B contain exactly the arcs between vertices that are at most k steps apart along the boundary. As no two vertices can be further apart than $\lceil \frac{n}{2} \rceil$ steps along the boundary, this implies that A eventually contains exactly the arcs of the visibility graph. More precisely, we inductively show that after step k of the iteration, $F[v_i]$ contains the vertices of $\text{chain}_{v_i}(v_{i+1}, v_{i+k})$ and $B[v_i]$ contains the vertices of $\text{chain}_{v_i}(v_{i-k}, v_{i-1})$ for all $v_i \in V$. For the sake of simplicity we abuse notation and write $F[v_i] = \text{chain}_{v_i}(v_{i+1}, v_{i+k})$ and $B[v_i] = \text{chain}_{v_i}(v_{i-k}, v_{i-1})$.

The discussion for $k = 1$ is trivial as every vertex has an arc to both its neighbors. The algorithm initializes F and B to consist of these arcs. It remains to show for all $0 \leq i < n$ that, assuming $F[v_i] = \text{chain}_{v_i}(v_{i+1}, v_{i+k-1})$ and $B[v_i] = \text{chain}_{v_i}(v_{i-k+1}, v_{i-1})$ after step $k-1$, we have $F[v_i] = \text{chain}_{v_i}(v_{i+1}, v_{i+k})$ and $B[v_i] = \text{chain}_{v_i}(v_{i-k}, v_{i-1})$ after step k.

The algorithm adds an arc between two vertices v_i and v_{i+k} if and only if there is a vertex $v_l \in \text{chain}(v_{i+1}, v_{i+k-1})$ with

$v_l \in F[v_i]$ and $v_l \in B[v_{i+k}]$ for which $\alpha_i + \alpha_j + \alpha_l = \pi$, where $\alpha_i, \alpha_j, \alpha_l$ are defined as in Algorithm 1. As v_i and v_l are less than k steps apart along the boundary, the induction assumption implies that $F[v_i] = \text{chain}_{v_i}(v_{i+1}, v_{i+k-1})$ and $B[v_{i+k}] = \text{chain}_{v_{i+k}}(v_{i+1}, v_{i+k-1})$. Therefore, $v_l \in F[v_i]$ and $v_l \in B[v_{i+k}]$ is equivalent to $(v_i, v_l), (v_{i+k}, v_l) \in A$ and, by Lemma 5.4, it suffices to show that $\alpha_i = \measuredangle^{\uparrow}_{v_i}(v_l, v_{i+k})\, , \alpha_j = \measuredangle^{\uparrow}_{v_{i+k}}(v_i, v_l)$ and $\alpha_l = \measuredangle_{v_l}(v_{i+k}, v_i)$ for all $v_l \in F[v_i] \cap B[v_{i+k}]$. By induction, we have $F[v_i] = \text{chain}_{v_i}(v_{i+1}, v_{i+k-1})$, and thus we have $\text{vis}_{F[v_i][v_l]}(v_i) = v_l$ and $\text{vis}_{|F[v_i]|+1}(v_i) = \arg\min_{v_b \in \text{chain}_{v_i}(v_{i+k}, v_{i-1})} \measuredangle_{v_i}(v_{i+1}, v_b)$. Consequently, we obtain that $\alpha_i = \angle_{v_i}(F[v_i]\,[v_l]\,, |F[v_i]| + 1) = \measuredangle^{\uparrow}_{v_i}(v_l, v_{i+k})$. Similarly, as v_l and v_{i+k} are less than k steps apart along the boundary, we get $\alpha_j = \measuredangle^{\uparrow}_{v_{i+k}}(v_i, v_l)$. By the induction assumption we also have $\text{vis}_{F[v_l][v_{i+k}]}(v_l) = v_{i+k}$ and $\text{vis}_{B[v_l][v_i]}(v_l) = v_i$ and thus $\alpha_l = \angle_{v_l}(F[v_l]\,[v_j]\,, B[v_l]\,[v_i]) = \measuredangle_{v_l}(v_{i+k}, v_i)$.

The uniqueness of the algorithm's solution follows from the fact that the existence of a triangle witness is necessary and sufficient for two vertices to see each other. □

Note that a polynomial running time is already achieved by a naïve implementation of the triangle witness algorithm.

Theorem 5.6. *An agent with boundary movement, angle sensor, and knowledge of n can solve the visibility graph reconstruction problem.*

Proof. We can simply define an exploration strategy that first collects all angle measurements while moving n times along the boundary. By Theorem 5.5, this information is sufficient to infer the visibility graph of the polygon that the agent is exploring. □

5.2. Reconstruction without Knowing n[†]

In the previous section we saw that an agent with angle sensor and knowledge of n can solve the visibility graph reconstruction problem, even if restricted to moving along the boundary only.

[†]The results presented in this section appeared in [20].

We will now show that in fact the agent does not need to know n a priori. Without this knowledge, there is no obvious way how to separate data collection and computation. Let G_{vis} denote the visibility graph that the agent is operating in. We propose an exploration strategy that computes in step j a subgraph of G_{vis} induced by a part of the boundary of length j. To be more precise, we will require some formalism.

We define the graph $G_i^j = (V_i^j, A_i^j)$ to be the subgraph of G_{vis} induced by $\{v_i, v_{i+1}, \ldots, v_j\}$. The degree of v_k in G_i^j is denoted by $d_i^j(v_k)$, with $d_0^{n-1}(v_k) = d(v_k)$. Observe that $G_{\text{vis}} = G_0^{n-1}$. By $\vec{\alpha}_k = (\alpha_{k,1}, \alpha_{k,2}, \ldots, \alpha_{k,d_k-1})$ we denote the sequence of angles at v_k, such that $\alpha_{k,x} = \measuredangle_{v_k}(x, x+1)$ is the angle between the x-th and $(x+1)$-th arc at v_k (in boundary order). Furthermore, we write

$$\measuredangle^{\uparrow}_{v_i}(v_j) := \sum_{x=1}^{d_i^{j-1}(v_i)} \alpha_{i,x}, \qquad \measuredangle^{\downarrow}_{v_j}(v_i) := \sum_{x=1}^{d_j - d_{i+1}^j(v_j)-1} \alpha_{j,x}$$

(cf. Figure 5.9). Note that the latter two quantities can be computed from $(G_i^{j-1}, \vec{\alpha}_i)$ and $(G_{i+1}^j, \vec{\alpha}_j)$, respectively. The following lemma relates these angles to the ones in the generalized angle-sum condition. Using this relation allows the agent check for a triangle witness of (v_i, v_j) as long as it has at its disposal G_i^{j-1}, , G_{i+1}^j, and $\vec{\alpha}_i, \vec{\alpha}_{i+1}, \ldots, \vec{\alpha}_j$.

Lemma 5.7. *For any vertices v_i, v_j and $v_l \in \text{chain}(v_{i+1}, v_{j-1})$, where v_l sees both v_i and v_j, we have*

$$\begin{aligned} \measuredangle^{\uparrow}_{v_i}(v_l, v_j) &= \measuredangle^{\uparrow}_{v_i}(v_j) - \measuredangle^{\uparrow}_{v_i}(v_l), \\ \measuredangle^{\uparrow}_{v_j}(v_i, v_l) &= \measuredangle^{\downarrow}_{v_j}(v_l) - \measuredangle^{\downarrow}_{v_j}(v_i), \\ \measuredangle_{v_l}(v_j, v_i) &= \measuredangle^{\downarrow}_{v_l}(v_i) - \measuredangle^{\uparrow}_{v_l}(v_j). \end{aligned}$$

Proof. Because v_i sees v_l, we have $\measuredangle^{\uparrow}_{v_i}(v_l, v_j) = \measuredangle_{v_i}(v_l, v_{j'})$, where $v_{j'}$ is the last vertex in $\text{chain}(v_{i+1}, v_j)$ visible to v_i. By definition, $\measuredangle^{\uparrow}_{v_i}(v_l) = \measuredangle_{v_i}(v_{i+1}, v_l)$ and $\measuredangle^{\uparrow}_{v_i}(v_j) = \measuredangle_{v_i}(v_{i+1}, v_{j'})$. The first line follows. The remaining lines can be shown analogously. □

Lemma 5.8. *The graph $G_i^j, 0 \le i < j$, can be computed from G_i^{j-1}, G_{i+1}^j, and $\vec{\alpha}_i, \vec{\alpha}_{i+1}, \ldots, \vec{\alpha}_j$.*

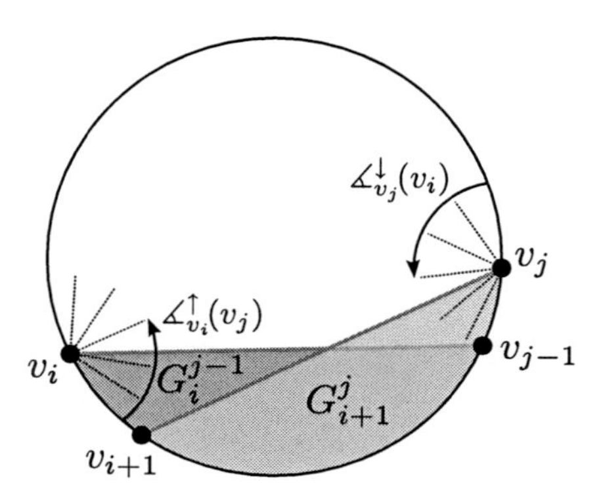

Figure 5.9.: Illustration of the angles $\measuredangle^{\uparrow}_{v_i}(v_j)$ and $\measuredangle^{\downarrow}_{v_j}(v_i)$.

Proof. Observe that A_i^j simply is the union of A_i^{j-1} and A_{i+1}^j, except for maybe the arcs (v_i, v_j) and (v_j, v_i). In order to compute G_i^j, it is hence sufficient to find out whether v_i sees v_j or not. If $j = i + 1$, this is always the case. Otherwise, by Lemma 5.4, v_i sees v_j if and only if there is a triangle witness of (v_i, v_j). As we know G_i^{j-1} and G_{i+1}^j, we can easily find all vertices $v_l \in \text{chain}(v_{i+1}, v_{j-1})$ that see both v_i and v_j. For each such vertex we can compute $\measuredangle^{\uparrow}_{v_i}(v_j)$, $\measuredangle^{\uparrow}_{v_i}(v_l)$, $\measuredangle^{\downarrow}_{v_j}(v_l)$, and $\measuredangle^{\downarrow}_{v_j}(v_i)$. By Lemma 5.7, these angles are sufficient in order to check the generalized angle-sum condition and thus to verify whether v_l is a triangle witness. Note that the entire procedure can be done in time polynomial in $j - i$. □

We can now give the exploration strategy that reconstructs the visibility graph (cf. Algorithm 2) and prove its correctness.

Theorem 5.9. *An agent with boundary movement and angle sensor can solve the visibility graph reconstruction problem.*

Proof. We prove that Algorithm 2 is correct and computes a unique solution. To this end, we claim that the exploration strategy maintains the invariant that in iteration j, the graphs G_x^y as well as $\vec{\alpha}_x$ have been determined for all $0 \leq x \leq y < j$. This is the case in the first iteration, because of the initialization in lines 1-3. In iteration j, the exploration strategy then computes G_x^j

Algorithm 2: Reconstructing G_{vis} without knowing n.

output: G_{vis}

1 $d_0 \leftarrow$ **degree**
2 $\vec{\alpha}_0 \leftarrow (\angle(1,2), \angle(2,3), \ldots, \angle(d_0 - 1, d_0))$
3 $G_0^0 \leftarrow (v_0, \emptyset)$
4 **for** $j \leftarrow 1, 2, \ldots$ **do**
5   **if** $d_0^{j-1}(v_0) = d_0$ **then**
6   **return** G_0^{j-1}
7   **move to** 1
8     $d_j \leftarrow$ **degree**
9     $\vec{\alpha}_j \leftarrow (\angle(1,2), \angle(2,3), \ldots, \angle(d_j - 1, d_j))$
10     $G_j^j \leftarrow (v_j, \emptyset)$
11   **for** $l \leftarrow j-1, j-2, \ldots, 1, 0$ **do**
12   compute G_l^j from $G_{l+1}^j, G_l^{j-1}, \vec{\alpha}_l, \vec{\alpha}_{l+1}, \ldots, \vec{\alpha}_j$

(using Lemma 5.8), for all $x \in [j-1]$, as well as $\vec{\alpha}_j$. Hence, the invariant is maintained.

The strategy terminates in the n-th iteration of the outer loop, since $d_0^n(v_0) = d(v_0) = d_0$. At that point, by the invariant, G_0^{n-1} has been determined. Correctness follows from $G_{\text{vis}} = G_0^{n-1}$. □

Observe that the agent performs exactly $n-1$ moves during the execution of Algorithm 2.

Part II.

Visibility Graph Exploration

Chapter 6.

Results for Strong Sensors*

In Chapter 5, we saw that an agent with boundary movement and angle sensor can solve the visibility graph reconstruction problem. Clearly, the angle-type sensor provides less information than the angle-sensor, and it remains an open problem whether an agent with boundary movement and angle-type sensor can solve the visibility graph reconstruction problem. Without movement restriction, we can show that an agent with angle-type sensor can solve the reconstruction problem, as long as it can also look back. In Chapters 8 and 9 we employ more advanced analysis to show that each of these capabilities on their own already suffices. We need the following lemmas.

Lemma 6.1. *Let u, v, w be three vertices of a polygon $\mathcal{P}$, such that v sees both u and w. Then,* $\mathrm{RE}(v,w) \cap \mathrm{LE}(u,v) = \mathrm{RE}(v,u) \cap \mathrm{LE}(w,v) = \mathrm{RE}(v,w) \cap \mathrm{RE}(v,u) = \mathrm{LE}(u,v) \cap \mathrm{LE}(w,v) = \emptyset$.

Proof. By Proposition 2.25, v does not see any vertex in $\mathrm{RE}(v,w)$, $\mathrm{RE}(v,u)$, $\mathrm{LE}(w,v)$, or $\mathrm{LE}(u,v)$. Since v sees both u and w, we get $u \notin \mathrm{RE}(v,w) \cup \mathrm{LE}(w,v)$ and $w \notin \mathrm{RE}(v,u) \cup \mathrm{LE}(u,v)$. Therefore, $\mathrm{RE}(v,w) \cap \mathrm{RE}(v,u) = \mathrm{LE}(u,v) \cap \mathrm{LE}(w,v) = \emptyset$.

Without loss of generality, assume $v \in \mathrm{chain}(u,w)$, and let x, y be the points where the rays $\overrightarrow{vw}, \overrightarrow{vu}$ first cross the boundary of $\mathcal{P}$, respectively. Since $v \in \mathrm{chain}(u,w)$, we have that x lies before y in boundary order starting at v ($x \neq y$, as $\mathcal{P}$ has no three collinear vertices). By definition of RE and LE, we have $\mathrm{RE}(v,w) \cap \mathrm{LE}(u,v) = \emptyset$. Because, by definition, $\mathrm{RE}(v,u)$ only contains vertices from $\mathrm{chain}(u,v)$ and $\mathrm{LE}(w,v)$ only contains vertices from $\mathrm{chain}(v,w)$, we get $\mathrm{RE}(v,u) \cap \mathrm{LE}(w,v) = \emptyset$. □

*The results presented in this chapter appeared in [7, 18].

Lemma 6.2. *Let u, v be two vertices of a polygon $\mathcal{P}$ and $A \in \{\mathrm{RE}(u,v), \mathrm{LE}(v,u)\}$. For every $w \in A$, exactly one of the following holds*

1. *v sees w,*
2. *there is exactly one vertex $v_b \in \mathrm{vis}(v)$, with $w \in \mathrm{RE}(v, v_b) \cup \mathrm{LE}(v_b, v)$.*

Proof. If v sees w, there cannot be a vertex $v_b \in \mathrm{vis}(v)$, such that $w \in \mathrm{RE}(v, v_b) \cup \mathrm{LE}(v_b, v)$, because such a vertex would be a blocker of (u, w) by Proposition 2.25.

Conversely, assume that there is a vertex $v_b \in \mathrm{vis}(v)$, such that $w \in \mathrm{RE}(v, v_b) \cup \mathrm{LE}(v_b, v)$. In this case, by Proposition 2.25, v does not see w. Also, there cannot be a second vertex $v_{b'} \in \mathrm{vis}(v) \setminus v_b$, such that $w \in \mathrm{RE}(v, v_{b'}) \cup \mathrm{LE}(v_{b'}, v)$, because

$$
\begin{aligned}
&(\mathrm{RE}(v, v_b) \cup \mathrm{LE}(v_b, v)) \cap (\mathrm{RE}(v, v_{b'}) \cup \mathrm{LE}(v_{b'}, v)) \\
&= (\mathrm{RE}(v, v_b) \cap \mathrm{RE}(v, v_{b'})) \cup (\mathrm{RE}(v, v_b) \cap \mathrm{LE}(v_{b'}, v)) \\
&\quad \cup (\mathrm{LE}(v_b, v) \cap \mathrm{RE}(v, v_{b'})) \cup (\mathrm{LE}(v_b, v) \cap \mathrm{LE}(v_{b'}, v)) \\
&= \emptyset,
\end{aligned}
$$

by Lemma 6.1. □

Theorem 6.3. *An agent with angle-type sensor and look-back sensor can solve the visibility graph reconstruction problem.*

Proof. We prove the theorem by presenting an exploration strategy for the agent to reconstruct the visibility graph of any polygon $\mathcal{P}$.

The agent moves from vertex to vertex along the boundary of $\mathcal{P}$ in boundary order. At each vertex v_i it iteratively identifies all visible vertices, i.e., it determines their global indices. The agent starts by identifying the vertex $\mathrm{vis}_1(v_i)$, which trivially has the global index $i + 1$. Further vertices can be identified using the following procedure. Observe that after each execution of the procedure, the agent is back at v_i.

Let $(u_1, u_2, \ldots, u_{d(v_i)}) := \mathrm{vis}(v_i)$. Now, let $v_k := u_x$ be the first vertex in $\mathrm{vis}(v_i)$ that has not yet been identified, and let $v_j := u_{x-1}$. In other words, the agent knows the index j and needs to find the index k. This is accomplished by counting all the vertices that are in $\mathrm{RE}(v_i, v_j)$ or $\mathrm{LE}(v_k, v_i)$. The total number N of these vertices simply needs to be added to $j+1$ in order to obtain $k = j + N + 1$. It remains to show how to determine N.

By Lemma 6.1, we have $\mathrm{RE}(v_i, v_j) \cap \mathrm{LE}(v_k, v_i) = \emptyset$. Therefore, $N = |\mathrm{RE}(v_i, v_j)| + |\mathrm{LE}(v_k, v_i)|$ and the agent can count the vertices in $\mathrm{RE}(v_i, v_j)$ and $\mathrm{LE}(v_k, v_i)$ separately. Algorithm 3 lists a returning exploration strategy for recursive counting. If v is the current location of the agent, the call "count(y, LE)" yields the number of vertices in $\mathrm{LE}(\mathrm{vis}_y(v_i), v_i)$, while the call "count($y$, RE)" yields the number of vertices in $\mathrm{RE}(v_i, \mathrm{vis}_y(v_i))$. After the execution of the strategy, the agent is guaranteed to be back at v. Using this method, we can determine N, because $N = \mathrm{count}(x-1, \mathrm{RE}) + \mathrm{count}(x, \mathrm{LE})$.

Algorithm 3: The method count

input : index y, parameter $o \in \{\mathrm{LE}, \mathrm{RE}\}$
output: if $o = \mathrm{LE}$: $|\mathrm{LE}(\mathrm{vis}_y(v), v)|$, else: $|\mathrm{RE}(v, \mathrm{vis}_y(v))|$

```
1  N ← 0
2  move to y
3  x ← look back
4  if o = LE then
5  |  A ← {x + 1, ..., d}
6  else
7  |  A ← {1, ..., x − 1}
8  foreach y ∈ A do
9  |  if ∠reflex(x, y) then
10 |  |  N ← N + 1 + count(y, LE) + count(y, RE)
11 move to i
```

The idea behind the counting method is that there are two types of vertices in $\mathrm{RE}(v, \mathrm{vis}_y(v))$ (respectively, in $\mathrm{LE}(\mathrm{vis}_y(v), v)$): vertices visible to $\mathrm{vis}_y(v)$ and vertices not visible to $\mathrm{vis}_y(v)$. For every of the former vertices, there is a reflex angle at $\mathrm{vis}_y(v)$, and by Lemma 6.2 the latter vertices can be counted recursively. It follows from Lemmas 6.1 and 6.2 that no vertex is counted twice during this procedure. □

Theorem 6.4. *An agent with angle sensor and compass can reconstruct every polygon $\mathcal{P}$.*

Proof. We are going to show that the agent can imitate an agent with angle-type and look-back sensor. By Theorem 6.3 this concludes the proof.

The angle sensor obviously yields enough information to emulate an angle-type sensor. It therefore suffices to show that the agent can imitate the capability of looking back. Assume the agent moves from a vertex v to a vertex u which it saw in global direction d when situated at v. From its new location u the agent knows that v lies in direction $-d$. Because, by definition, no three vertices of a polygon are collinear, the agent is guaranteed to see no other vertices in that direction, and thus the agent is capable of uniquely identifying the vertex it came from. In other words, the agent is capable of looking back. □

Chapter 7.

General Tools for Mapping

This chapter revolves around structural properties of the minimum base graph $G^\star_{\text{vis}}$ of the arc-labeled visibility graph G_{vis} induced by an agent exploring some polygon $\mathcal{P}$. By definition of the minimum base graph, $G^\star_{\text{vis}}$ encodes all the data that the agent can collect while exploring $\mathcal{P}$. Intuitively, the knowledge of $G^\star_{\text{vis}}$ allows to simulate the moves and observations made by the agent. Therefore, once the agent knows $G^\star_{\text{vis}}$ together with the vertex that corresponds to its current location, it does not need to make any more moves or observations in $\mathcal{P}$ directly. This suggests a natural way of dividing the question whether the agent can solve the reconstruction problem into two parts:

1. Can the agent infer $G^\star_{\text{vis}}$?
2. Is the information encoded in $G^\star_{\text{vis}}$ sufficient in order to solve the reconstruction problem?

We will show that an agent that knows an upper bound $\bar{n}$ on the number of vertices of G_{vis} can always infer $G^\star_{\text{vis}}$. What is even more, our proof will not exploit the special structure of visibility graphs, and thus even yield the stronger result that an agent knowing $\bar{n}$ can find the minimum base graph of all arc-labeled graphs of size at most $\bar{n}$.

Concerning the second question, we cannot hope for a positive answer in general, since there are pairs of non-isomorphic arc-labeled graphs that share the same minimum base graph (cf. Figure 2.4). We will see that under certain conditions, the nature of visibility graphs allows to expose a certain structure of the minimum base graph of an arc-labeled visibility graph. This structure will prove useful for solving the weak rendezvous problem and also as a basis for reconstruction algorithms in later chapters.

7.1. Finding the Minimum Base Graph*

In this section we consider the general problem of exploring a general arc-labeled graph $G = (V, A, \lambda)$ with an agent. The only extension to the agent model that we make is to give the agent knowledge of an upper bound $\bar{n}$ on the total number of vertices n. We let $\mathscr{G}_{\bar{n}} \subset \mathscr{G}$ denote the family of all arc-labeled graphs which encode an upper bound on their size in the arc-labeling. We first show that any two prime graphs can be distinguished using a label-sequence of finite length. This result is mainly due to the graphs being locally oriented.

Lemma 7.1. *Let $G_1 = (V_1, A_1, \lambda_1)$, $G_2 = (V_2, A_2, \lambda_2)$ be two distinct prime graphs. There is a label-sequence δ for which $\delta(G_1) \neq \emptyset$ and $\delta(G_2) = \emptyset$, or vice versa.*

Proof. First observe that for every pair of vertices $u_1 \in V_1, u_2 \in V_2$ there is a label-sequence δ_{u_1,u_2} for which $\delta_{u_1,u_2}(u_1) \neq \emptyset$ and $\delta_{u_1,u_2}(u_2) = \emptyset$, or vice versa. This follows because, by definition, u_1 and u_2 cannot be indistinguishable. Hence, by Proposition 2.48, a label-sequence δ_{u_1,u_2} exists, with $\delta_{u_1,u_2}(u_1) \neq \emptyset$ and $\delta_{u_1,u_2}(u_2) = \emptyset$, or vice versa.

We now describe how to obtain the desired label-sequence δ. We start with the empty label-sequence $\delta^{(0)}$ and iteratively extend it to a longer but still finite label-sequence $\delta^{(i)}$ in step i. Let $U_l^{(i)} := \{v \in V_l | \delta^{(i)}(v) \neq \emptyset\}$, $l \in [2]$, be the sets of vertices of G_1 and G_2, respectively, that are "compatible" with $\delta^{(i)}$. As $\delta^{(i+1)}$ extends $\delta^{(i)}$, we have by construction that $U_l^{(i+1)} \subseteq U_l^{(i)}$ for $l \in [2]$. We show that our extension satisfies $\left(U_1^{(i+1)} \cup U_2^{(i+1)}\right) \subsetneq \left(U_1^{(i)} \cup U_2^{(i)}\right)$ in every step and that either $\delta^{(i+1)}(G_1) \neq \emptyset$ or $\delta^{(i+1)}(G_2) \neq \emptyset$. At some point we thus obtain a label-sequence δ for which exactly one of the two graphs has no compatible vertices. It remains to show the existence of an appropriate extension.

Let $\delta^{(i)}$ be a finite label-sequence with $\delta^{(i)}(G_1) \neq \emptyset$ or $\delta^{(i)}(G_2) \neq \emptyset$. If $U_l^{(i)} = \emptyset$ for one $l \in [2]$, we have either $\delta^{(i)}(G_1) = \emptyset$ or

*The results presented in this section appeared in [11].

$\delta^{(i)}(G_2) = \emptyset$. We can thus set $\delta = \delta^{(i)}$ and we are done. So assume $U_l^{(i)} \neq \emptyset$ for $l \in [2]$. Then, there are two vertices $w_l \in U_l^{(i)}$, $l \in [2]$. Let $p_l = \delta^{(i)}(w_l)$, and $u_l := \text{target}(p_l)$ for $l \in [2]$. By our initial observation, there is a label-sequence δ_{u_1,u_2} with $\delta_{u_1,u_2}(u_1) \neq \emptyset$ and $\delta_{u_1,u_2}(u_2) = \emptyset$, or vice versa. We set $\delta^{(i+1)} = \delta^{(i)} \circ \delta_{u_1,u_2}$. By definition of δ_{u_1,u_2}, we have $\delta^{(i+1)}(a) \neq \emptyset$ and $\delta^{(i+1)}(b) = \emptyset$, or vice versa. Thus $U_1^{(i+1)} \subsetneq U_1^{(i)}$ and $\delta^{(i+1)}(G_2) \neq \emptyset$, or $U_2^{(i+1)} \subsetneq U_2^{(i)}$ and $\delta^{(i+1)}(G_1) \neq \emptyset$. □

Corollary 7.2. *Let $G_1 = (V_1, A_1, \lambda_1), G_2 = (V_2, A_2, \lambda_2)$ be two distinct prime graphs, and $u_1 \in V_1, u_2 \in V_2$. There is a label-sequence δ for which $\delta(u_1) \neq \emptyset$ and $\delta(G_2) = \emptyset$, or $\delta(u_2) \neq \emptyset$ and $\delta(G_1) = \emptyset$.*

Proof. By Lemma 7.1, there is a label-sequence δ' for which $\delta'(G_1) \neq \emptyset$ and $\delta'(G_2) = \emptyset$, or vice versa. Assume $\delta'(G_1) \neq \emptyset$ and let $w_1 \in V_1$ with $\delta'(w_1) \neq \emptyset$. Let p be a (possibly trivial) path from u_1 to w_1 and $\Pi = \lambda(p)$ be the associated label-sequence. We set $\delta := \Pi \circ \delta'$ and obtain a label-sequence for which $\delta(u_1) \neq \emptyset$ and $\delta(G_2) = \emptyset$. If instead $\delta'(G_2) \neq \emptyset$, an analogous argument yields a label-sequence δ for which $\delta(u_2) \neq \emptyset$ and $\delta(G_1) = \emptyset$. □

The following theorems give a general method for an agent exploring an arc-labeled graph G to determine the minimum base graph $G^\star$ of G and the agent's position therein. Note that while the described exploration strategies are applicable in general graphs, they are computationally expensive and require an asymptotically exponential number of moves in the size of the graph. For visibility graphs with particular arc-labelings that encode geometrical properties, we can hope for more efficient ways of finding the minimum base graph.

Theorem 7.3. *There is an exploration strategy that computes the minimum base graph $G^\star$ in every graph $G \in \mathscr{G}_{\bar{n}}$.*

Proof. By Theorem 2.50, $G^\star$ is unique. We will give an exploration strategy that maintains a finite set $C \subseteq \mathscr{G}_{\bar{n}}$ of graphs for which it guarantees $G^\star \in C$ at all times. We begin by setting C to contain all prime graphs of size at most $\bar{n}$. While $|C| > 1$, we let G_1, G_2 be two graphs in C and describe how to conclude that either G_1 or G_2 can be eliminated from C. Once $|C| = 1$, the only graph left will have to be $G^\star$. In the following, let p_{hist} denote the walk in G that the agent has travelled along so far during the execution of the exploration strategy, and let $\Lambda_{\text{hist}} = \lambda(p_{\text{hist}})$ be the associated label-sequence. Note that the agent is aware of Λ_{hist} but not of p_{hist}, since it does not know the graph G nor its starting location in G. We use $v_{\text{hyp}}(v_{\text{start}})$ to denote the last vertex on the walk $\Lambda_{\text{hist}}(v_{\text{start}})$, i.e., the vertex the agent would currently be located at if it had started the exploration at vertex v_{start}.

Given two prime graphs $G_1 = (V_1, A_1), G_2 = (V_2, A_2)$, we argue how to conclude that one of them cannot be $G^\star$. Let $\mathcal{V}_1, \mathcal{V}_2$ denote the vertices of G_1 and G_2, respectively, at which the agent could possibly have started the exploration. We begin by setting $\mathcal{V}_l = \{v \in V_l | \, \Lambda_{\text{hist}}(v) \neq \emptyset\}$ for $l \in [2]$. Let $u_1 \in \mathcal{V}_1, u_2 \in \mathcal{V}_2$. By Corollary 7.2, there is a label-sequence δ for which $\delta(v_{\text{hyp}}(u_1)) \neq \emptyset$ and $\delta(G_2) = \emptyset$ or $\delta(v_{\text{hyp}}(u_2)) \neq \emptyset$ and $\delta(G_1) = \emptyset$. The agent can compute such a sequence δ simply by checking all possible label-sequences in order of increasing lengths. The agent tries to move along a path corresponding to δ. If that turns out not to be possible because at some point no arc has the required label, it can discard a from $\mathcal{V}_1$ if $\delta(v_{\text{hyp}}(u_1)) \neq \emptyset$, and u_2 from $\mathcal{V}_2$ otherwise. If it successfully traced δ, the agent can eliminate G_1 from C if $\delta(G_1) = \emptyset$, and G_2 otherwise. As long as neither G_1 nor G_2 is discarded, the agent continues with a new choice of $u_1 \in \mathcal{V}_1, u_2 \in \mathcal{V}_2$. Once $\mathcal{V}_1$ or $\mathcal{V}_2$ is exhausted, the corresponding graph G_1 or G_2, respectively, can be eliminated. □

The following theorem states that an agent can move to any given vertex of a prime graph G without knowing its starting location. Afterwards, the agent can maintain knowledge of its location, since it can keep track of all moves in G.

Theorem 7.4. *Let $\mathscr{G}'_{\bar{n}} \subseteq \mathscr{G}_{\bar{n}}$ contain only prime graphs. There is an exploration strategy that moves an agent exploring any given graph $G \in \mathscr{G}'_{\bar{n}}$ to a specified vertex v of G.*

Proof. By definition of prime graphs, no two vertices $u \neq w$ of G are indistinguishable. By Proposition 2.48, this means that for every two vertices $u \neq w$ of G, there is a label-sequence $\delta_{u,w}$

with $\delta_{u,w}(u) \neq \emptyset$ and $\delta_{u,w}(w) = \emptyset$, or vice versa. We now give an exploration strategy that maintains a set $\mathcal{V} \subseteq V$ of possible locations at which the agent could be located according to its observations so far. In the beginning, we set $\mathcal{V} = V$ since the agent has not made any observations yet. In each step, the agent reduces the size of $\mathcal{V}$, so that in the end only its actual location remains. We will now show how to perform this reduction.

While $|\mathcal{V}| \geq 2$, let $u, w \in \mathcal{V}$ be two distinct candidates for the agent's location. The agent tries to move according to the label-sequence $\delta_{u,w}$. If this is possible, the agent updates $\mathcal{V}$ to the new set $\mathcal{V}' = \{\text{target}(\delta_{u,w}(v)) | v \in \mathcal{V} \wedge \delta_{u,w}(v) \neq \emptyset\}$. We have $|\mathcal{V}'| < |\mathcal{V}|$, since $\delta_{u,w}(u) = \emptyset$ or $\delta_{u,w}(w) = \emptyset$. If the agent could not move according to $\delta_{u,w}$, let δ denote the prefix of $\delta_{u,w}$ that the agent could successfully trace. It updates $\mathcal{V}$ to the new set $\mathcal{V}' = \{\text{target}(\delta(v)) | v \in \mathcal{V} \wedge \delta_{u,w}(v) = \emptyset \wedge \delta(v) \neq \emptyset\}$. We have $|\mathcal{V}'| < |\mathcal{V}|$, since $\delta_{u,w}(u) \neq \emptyset$ or $\delta_{u,w}(w) \neq \emptyset$.

Once $|\mathcal{V}| = 1$, the agent is located at the last vertex remaining in $\mathcal{V}$. All that remains is to move from this vertex to v, which is possible as G is known to the agent and strongly connected. □

Corollary 7.5. *There is an exploration strategy for $\mathscr{G}_{\bar{n}}$ that moves an agent exploring any graph $G \in \mathscr{G}_{\bar{n}}$ to an arbitrary vertex v of G and computes $G^\star$ as well as the identity of the vertex representing C_v in $G^\star$.*

Proof. By Theorem 7.3, the agent can determine $G^\star$. Afterwards, the agent can pretend to be exploring $G^\star$ instead of G, since G and $G^\star$ are indistinguishable. By Theorem 7.4, the agent can move to any vertex $v^\star$ of $G^\star$ that it chooses. In G, the agent will then be located at a vertex of $C_{v^\star}$. □

7.2. Visibility Graphs with Optimal Substructure

We now come back to the exploration of visibility graphs. We expose a particular structure which is present in many cases and allows us to derive useful properties.

encode ears

Definition 7.6. We say an arc-labeled visibility graph G_{vis} *encodes ears* if for every ear e of G_{vis}, all vertices in the class C_e are ears.

structural subgraph

Definition 7.7. Let $G = (V, A, \lambda), G' = (V', A', \lambda')$ be arc-labeled visibility graphs. We say G' is a *structural subgraph* of G if (V', A') is an induced subgraph of (V, A).

local substructure

Definition 7.8. We say $G_{\text{vis}} \in \mathscr{F}$ has *local substructure* with respect to a family $\mathscr{F}'$ if for all structural subgraphs $G'_{\text{vis}} \in \mathscr{F}'$ of G_{vis}, the arc-label of any arc (u, v) of G'_{vis} can be inferred from the arc-label of the arc (u, v) in G_{vis} and the arc-labels in G_{vis} of the arcs at u and v in G'_{vis}.

optimal substructure

Definition 7.9. We say $G_{\text{vis}} \in \mathscr{F}$ has *optimal substructure* if there exists a complete family $\mathscr{F}' \subset \mathscr{F}$, $G_{\text{vis}} \in \mathscr{F}'$, such that every structural subgraph $G'_{\text{vis}} \in \mathscr{F}'$ of G_{vis} encodes ears and has local substructure with respect to $\mathscr{F}'$.

admit optimal substructure

Definition 7.10. Let $\mathscr{F}' \subset \mathscr{F}$ be a family that encodes some agent model. We say the agent model *admits optimal substructure* if every graph in $\mathscr{F}'$ has optimal substructure.

Theorem 7.11. *Every $G_{\text{vis}} = (V, A, \lambda) \in \mathscr{F}$ with optimal substructure has a class that forms a clique.*

Proof. First note that, by Proposition 2.19, G_{vis} has an ear e. We prove the theorem by induction on the number k of classes of G_{vis}. For $k = 1$, every vertex of G_{vis} is an ear, since G_{vis} encodes ears. Therefore, every vertex of G_{vis} must be convex. By Proposition 2.37 this means that there cannot be a vertex on the interior of any Euclidean shortest path in any polygon $\mathcal{P}$ with visibility graph G_{vis}. In other words, all vertices of G_{vis} see each other, i.e., G_{vis} is a clique.

Now assume $k > 1$. Let $\mathscr{F}' \subset \mathscr{F}$ be as in Definition 7.9. By Proposition 2.20, we can cut all the vertices of C_e off $\mathcal{P}$, obtaining a smaller polygon $\mathcal{P}'$. Let $G'_{\text{vis}} = (V', A', \lambda') \in \mathscr{F}'$ be the arc-labeled visibility graph of $\mathcal{P}'$ in $\mathscr{F}'$. We claim that two vertices $u, w \in V'$ that are in the same class in G_{vis} are also in the same class in G'_{vis}. Once we established this claim, it follows that G'_{vis} must have fewer classes than G_{vis}, and hence, by induction, G'_{vis}

has at least one class that forms a clique. Since the vertices in any class of G_{vis} except for C_e must be contained in a single class of G'_{vis}, it follows that at least one class of G_{vis} forms a clique in G_{vis}. It remains to prove the claim.

Let $G^-_{\text{vis}} := (V', A', \lambda|_{A'})$, where $\lambda|_{A'}$ denotes the restriction of λ to A'. Let $G^\star_{\text{vis}} = (V^\star, A^\star, \lambda^\star)$ be the minimum base graph of G_{vis}, and let $G^{\star-}_{\text{vis}} = (V^{\star-}, A^{\star-}, \lambda^\star|_{A^{\star-}})$ be the graph obtained by removing the vertex representing the class C_e of G_{vis} from $G^\star_{\text{vis}}$. As we removed all vertices of class C_e, we have that G^-_{vis} is indistinguishable from $G^{\star-}_{\text{vis}}$. Let $\mathcal{L}_G(v)$ denote the set of arc labels of all arcs at node v in graph G. Because G_{vis} has local substructure with respect to $\mathscr{F}'$, there is a function φ such that for every arc $(a,b) \in A'$ we have $\lambda'(a,b) = \varphi(\lambda(a,b)\,,\ \mathcal{L}_{G^-_{\text{vis}}}(a)\,, \mathcal{L}_{G^-_{\text{vis}}}(b))$. Because $G^\star_{\text{vis}}, G^{\star-}_{\text{vis}}$ are indistinguishable from $G_{\text{vis}}, G^-_{\text{vis}}$, respectively, it follows that for every arc $(a^\star, b^\star) \in A^{\star-}$ there must be an arc $(a,b) \in A^-$ with $\mathcal{L}_{G^{\star-}_{\text{vis}}}(a^\star) = \mathcal{L}_{G^-_{\text{vis}}}(a)$ and $\mathcal{L}_{G^{\star-}_{\text{vis}}}(b^\star) = \mathcal{L}_{G^-_{\text{vis}}}(b)$. We can therefore define $\lambda^{\star-\prime}$ by $\lambda^{\star-\prime}(a^\star, b^\star) = \varphi(\lambda^\star(a^\star, b^\star)\,, \mathcal{L}_{G^{\star-}_{\text{vis}}}(a^\star)\,, \mathcal{L}_{G^{\star-}_{\text{vis}}}(b^\star))$ for every arc $(a^\star, b^\star) \in A^{\star-}$. Let $(a^\star, b^\star)$ be an arc of $G^{\star-}_{\text{vis}}$. Then, for every arc (a,b) of G^-_{vis} that corresponds to the arc $(a^\star, b^\star)$, by definition of φ, we have $\lambda(a,b) = \lambda^\star(a^\star, b^\star)$ if and only if $\lambda'(a,b) = \lambda^{\star-\prime}(a^\star, b^\star)$, since $\mathcal{L}_{G^{\star-}_{\text{vis}}}(a^\star) = \mathcal{L}_{G^-_{\text{vis}}}(a)$ and $\mathcal{L}_{G^{\star-}_{\text{vis}}}(b^\star) = \mathcal{L}_{G^-_{\text{vis}}}(b)$. Hence, G'_{vis} is indistinguishable from $G^{\star-\prime}_{\text{vis}} := (V^{\star-}, A^{\star-}, \lambda^{\star-\prime})$, since G^-_{vis} was indistinguishable from $G^{\star-}_{\text{vis}}$. This means that G'_{vis} and $G^{\star-\prime}_{\text{vis}}$ share the same minimum base graph $G'^\star_{\text{vis}}$. Therefore, all vertices in G'_{vis} that are indistinguishable from a single vertex in $G^{\star-\prime}_{\text{vis}}$ lie in the same class. Because u, w are in the same class of G_{vis}, in G_{vis} they must be indistinguishable from the same vertex $v^\star$ in $G^\star_{\text{vis}}$. But then, in G'_{vis}, they must be indistinguishable from $v^\star$ in $G^{\star-\prime}_{\text{vis}}$. We conclude that u and w are in the same class in G'_{vis} and we established the claim. □

Corollary 7.12. *The size of a visibility graph $G_{\text{vis}} \in \mathscr{F}$ with optimal substructure can be inferred from $G^\star_{\text{vis}}$.*

Proof. By Theorem 7.11, G_{vis} has a class that forms a clique. Let q be the number of self-loops of the corresponding vertex in $G^\star_{\text{vis}}$. By Proposition 2.57, we can compute n via $n = (q+1) \cdot n^\star$, where $n^\star$ is the size of $G^\star_{\text{vis}}$. □

7.3. Solving the Weak Rendezvous Problem

In this section, we show that optimal substructure allows multiple agents to agree on a class of the arc-labeled visibility graph and meet there. To accomplish this, agents need to have a way to distinguish between the classes of an arc-labeled visibility graph. In the following, we establish formal means for this purpose.

$\delta^{<}_{u,w}$

Lemma 7.13. *For every two vertices u, w of a prime graph G there is a label-sequence $\delta_{u,w}$ such that $\delta_{u,w}(u) \neq \emptyset$ and $\delta_{u,w}(w) = \emptyset$, or vice versa. By $\delta^{<}_{u,w}$ we denote the lexicographically smallest such sequence.*

Proof. No two vertices of a prime graph are indistinguishable. Hence, by Proposition 2.48, for every two vertices $u, w \in V$ there is a label-sequence $\delta_{u,w}$ with $\delta_{u,w}(u) \neq \emptyset$ and $\delta_{u,w}(w) = \emptyset$, or vice versa. □

characteristic tuple

Definition 7.14. Let u be a vertex of a prime graph $G = (V, A, \lambda)$, and define two sets $S := \{ \delta^{<}_{u,w} \mid w \in V \wedge \delta^{<}_{u,w}(u) \neq \emptyset \}$, $\bar{S} := \{ \delta^{<}_{u,w} \mid w \in V \wedge \delta^{<}_{u,w}(u) = \emptyset \}$. The *characteristic tuple* of u is the tuple $(\mathcal{S}, \bar{\mathcal{S}})$, where $\mathcal{S}, \bar{\mathcal{S}}$ are the sequences containing the elements of $S, \bar{S}$, respectively, in lexicographical order.

Lemma 7.15. *No two vertices of a prime graph G have the same characteristic tuple.*

Proof. Consider any two vertices u, w of G, and let their characteristic tuples be $(\mathcal{A}_1, \mathcal{B}_1), (\mathcal{A}_2, \mathcal{B}_2)$, respectively. By definition, $\delta^{<}_{u,w}$ is either in $\mathcal{A}_1$ or in $\mathcal{A}_2$, but not in both. □

$<$

Definition 7.16. Let $G_{\text{vis}} \in \mathscr{F}$ and $u^\star, w^\star$ be two vertices of $G^\star_{\text{vis}}$. We write $C_{u^\star} < C_{w^\star}$ if the characteristic tuple of $u^\star$ is lexicographically smaller than the one of $w^\star$.

Lemma 7.17. *The relation '$<$' is a total order on the classes of a visibility graph.*

Proof. The claim follows from the definition of '$<$' together with Lemma 7.15. □

$C^\star$

Definition 7.18. Let $G_{\text{vis}} \in \mathscr{F}$ have optimal substructure. By $C^\star$ we denote the smallest class of G_{vis}, with respect to '$<$', that forms a clique in G_{vis}.

We can now give criteria for when agents can solve the strong and the weak rendezvous problem, respectively.

Theorem 7.19. *Any number $k > 1$ of agents can solve the strong rendezvous problem in $G \in \mathscr{G}_{\bar{n}}$ if and only if G is a prime graph.*

Proof. The first part of the proof follows from Proposition 2.61. For the converse, it suffices to show that in a prime graph, an agent can move to the smallest class with respect to '$<$'. If all agents do this, they meet at the same vertex since every class has size one.

By Theorem 7.3, and because G is a prime graph, the agent can infer G. For every pair of vertices u, v of G, the agent can find $\delta_{u,w}$ simply by trying all possible label-sequences in order of increasing lengths and checking whether they have the desired property by inspecting G. Therefore, the agent can infer the characteristic tuples of each vertex of G, and thus identify the smallest vertex $v^<$ with respect to '$<$'. By Theorem 7.4, the agent can continue moving, until it knows its current location. Finally, since G is strongly connected, the agent can then move to $v^<$ by tracing a path from its current location in G. □

Let $\mathscr{F}_{\bar{n}} \subseteq \mathscr{G}_{\bar{n}}$ denote the family of all arc-labeled visibility graphs in $\mathscr{G}_{\bar{n}}$.

Theorem 7.20. *There is an exploration strategy for $\mathscr{F}_{\bar{n}}$ that moves an agent exploring any graph $G_{\text{vis}} \in \mathscr{F}_{\bar{n}}$ with optimal substructure to a vertex of class $C^\star$.*

Proof. By Theorem 7.3, the agent can infer $G^\star_{\text{vis}}$. By Theorem 7.11, G_{vis} has at least one class that forms a clique. The agent can identify all those classes simply by inspecting the number of self-loops of every vertex of $G^\star_{\text{vis}}$. For every pair of vertices $u^\star, v^\star$ of $G^\star_{\text{vis}}$, the agent can find $\delta_{u^\star,w^\star}$ simply by trying all possible label-sequences in order of increasing lengths and checking whether they have the desired property by inspecting $G^\star_{\text{vis}}$.

Therefore, the agent can infer the characteristic tuples of each vertex of $G^{\star}_{\text{vis}}$, and thus identify the class $C^{\star}$. Finally, by Corollary 7.5, the agent can continue moving, until it knows the class of its current location. Since G_{vis} is strongly connected, the agent can then move to a vertex of $C^{\star}$ by tracing a path in $G^{\star}_{\text{vis}}$ to the vertex corresponding to $C^{\star}$. □

Theorem 7.21. *If an agent model admits optimal substructure, then any number of agents with knowledge of $\bar{n}$ can solve the weak rendezvous problem.*

Proof. By Theorem 7.20, every agent can position itself on a vertex of $C^{\star}$. Once every agent has done this, all agents mutually see each other, since, by definition, $C^{\star}$ is a clique. □

Chapter 8.

Look-Back Sensor*

In this chapter we consider agents with look-back sensor and knowledge of an upper bound on the number of vertices. By $\mathscr{F}_{\mathrm{LB}} \subseteq \mathscr{F}_{\bar{n}}$ we denote a family of arc-labeled visibility graphs that encodes this agent model.

We show that there is a one-to-one mapping between the graphs in $\mathscr{F}_{\mathrm{LB}}$ and their minimum base graphs. With the results from Section 7.1 this implies that an agent with look-back sensor and knowledge of an upper bound on the number of vertices can solve the visibility graph reconstruction problem. We then give an alternate strategy for such agents to infer the minimum base graph of the graph they are exploring, which leads to a polynomial exploration strategy for the visibility graph reconstruction problem.

In Section 2.4.1, we described how to encode the information available through the look-back sensor in the arc-labels (cf. Figure 2.11). Knowledge of an upper bound $\bar{n}$ on the number of vertices can be reflected by appending $\bar{n}$ to every arc-label. We make this precise and fix the arc-label of an arc (u, v) to be $(a, b, \bar{n})$, where a is the index of (u, v) in the local ordering at u, b is the index of (v, u) in the local ordering at v. We define $\mathrm{LB}_u(a) := b$ to be the second element of the arc label $(a, b, \bar{n})$ of the arc (u, v). By $-X$ we denote the index $d(v) - X + 1$ for arcs at v. A boundary arc has label $(1, -1, \bar{n})$ or $(-1, 1, \bar{n})$, depending on whether it leads to the next or previous vertex along the boundary.

*The results presented in this chapter appeared in [10, 12].

8.1. Reconstruction with the General Method

By Theorem 7.1, the agent can can infer the minimum base graph of the visibility graph it is exploring. In order to show that the agent can solve the reconstruction problem, we need to prove that there is a one-to-one mapping between the graphs in $\mathscr{F}_{\mathrm{LB}}$ and their minimum base graphs. We will first show that the agent model admits optimal substructure by showing that every graph in $\mathscr{F}_{\mathrm{LB}}$ has optimal substructure.

Lemma 8.1. *Let $G_{\mathrm{vis}} \in \mathscr{F}_{\mathrm{LB}}$ and define the label-sequence $\Lambda := ((1,-1,\bar{n}),(-2,2,\bar{n}))$, where $\bar{n}$ is the bound on the size of G_{vis} as encoded in its arc-labeling. A vertex v_i of G_{vis} is an ear if and only if $\Lambda(v_i) \neq \emptyset$.*

Proof. If v_i is an ear, then v_{i+1} and v_{i-1} see each other. In particular, the arc (v_{i+1}, v_{i-1}) has index $-2 \equiv d(v_{i+1}) - 1$ in the local ordering at v_{i+1}. Similarly, the arc (v_{i-1}, v_{i+1}) has index 2 in the local ordering at v_{i-1}. Therefore, $\Lambda(v_i) = v_{i-1}$.

Now let v_i be a vertex with $\Lambda(v_i) \neq \emptyset$. We then have $\Lambda'(v_{i+1}) \neq \emptyset$, where $\Lambda' := ((d(v_{i+1}) - 1, 2, \bar{n}))$. For the sake of contradiction, assume $\mathrm{vis}_{-2}(v_{i+1}) \neq v_{i-1}$. Consider the ordered cycle formed by $\mathrm{chain}(\mathrm{vis}_{-2}(v_{i+1}), v_{i+1})$. Because $\mathrm{vis}_{-2}(v_{i+1}) \neq v_{i-1}$, this cycle has length at least four. In the subpolygon it induces, the vertices v_{i+1} and $\mathrm{vis}_{-2}(v_{i+1})$ have degree two. This is a contradiction with Proposition 2.34. □

Lemma 8.2. *Every graph in $\mathscr{F}_{\mathrm{LB}}$ has optimal substructure.*

Proof. Let $\mathscr{F}'_{\mathrm{LB}} \subset \mathscr{F}$ be the complete family that encodes the look back sensor and knowledge of $\bar{n}$ as described before. By definition, it is sufficient to show that every graph $G_{\mathrm{vis}} \in \mathscr{F}_{\mathrm{LB}}$ encodes ears and has local substructure with respect to $\mathscr{F}'_{\mathrm{LB}}$.

Let v_i be an ear of G_{vis} and let Λ be defined as in Lemma 8.1. By Lemma 8.1, we have $\Lambda(v_i) \neq \emptyset$. Since all vertices in C_{v_i} are indistinguishable from v_i, by Proposition 2.48, $\Lambda(v) \neq \emptyset$ for every $v \in C_{v_i}$. Applying Lemma 8.1 in the other direction yields that every vertex of C_{v_i} is an ear. Hence, G_{vis} encodes ears.

It is easy to see that G_{vis} has local substructure with respect to $\mathscr{F}'_{\text{LB}}$, since the relative order of the arcs at every vertex stays the same in every induced subgraph. To obtain the new label of an arc in the induced subgraph, its order and look-back information in G_{vis}, which is encoded in its label, simply needs to be modified by shifting all indices by the number of arcs of lower index that fell away at its target and source vertices. □

Theorem 8.3. *Any number of agents with look back sensor and knowledge of $\bar{n}$ can solve the weak rendezvous problem.*

Proof. The statement is a direct consequence of Theorem 7.21 combined with Lemma 8.2. □

Corollary 8.4. *Every graph in $\mathscr{F}_{\text{LB}}$ has a class that forms a clique.*

Proof. The statement is a direct consequence of Theorem 7.11 combined with Lemma 8.2. □

The remainder of this section makes use of Corollary 8.4 to show that every graph in $\mathscr{F}_{\text{LB}}$ can be computed from its minimum base graph.

Theorem 8.5. *Every graph in $\mathscr{F}_{\text{LB}}$ can be determined from its minimum base graph in polynomial time.*

Proof. We describe an algorithm to determine any graph $G_{\text{vis}} \in \mathscr{F}_{\text{LB}}$, given $G^\star_{\text{vis}}$. Let $n^\star$ denote the size of $G^\star_{\text{vis}}$. By Corollary 7.12, we can determine the size n of G_{vis} from $G^\star_{\text{vis}}$. As usual we use $v_0, v_1, \ldots, v_{n-1}$ to refer to the vertices of G_{vis} in boundary order. By Proposition 2.56, classes repeat periodically along the boundary, hence we know how to group the vertices into classes. We are free to fix any vertex $v^\star_0$ of $G^\star_{\text{vis}}$ and set $C_{v^\star_0} = C_{v_0}$. Because the arc-labeling of $G^\star_{\text{vis}}$, by definition, encodes the order of the arcs at every vertex, this fixes the relationship between the vertices of $G^\star_{\text{vis}}$ and G_{vis}, i.e., for each vertex v of G_{vis} we know the vertex $v^\star$ of $G^\star_{\text{vis}}$ such that $C_v = C_{v^\star}$. It remains to identify the arcs of G_{vis} – their labels can then trivially be inferred from $G^\star_{\text{vis}}$. In the following, let $(C_0, C_1, \ldots, C_{n-1}) := \left(C_{v^\star_0}, C_{v_1}, \ldots, C_{v_{n-1}}\right)$ denote the sequence of classes around the boundary.

By Corollary 8.4, G_{vis} has at least one class that forms a clique. Every such class can be recognized in $G^{\star}_{\text{vis}}$ as a vertex with a maximum number of self-loops, since all classes of G_{vis} have the same size (Proposition 2.57). Without loss of generality, assume C_0 is a clique in G_{vis}. Let $v_i \in C_0$ be a vertex of this class. We argue that $\text{vis}_k(v_i)$, and hence the k-th arc at v_i, is easily determined for every $k \in [d(v_i)]$. From $G^{\star}_{\text{vis}}$ we know the class C with $\text{vis}_k(v_i) \in C$. Let $x_k := |\{l \in [k-1] \,|\, \text{vis}_l(v_i) \in C_0\}|$ be the number of arcs with index smaller k that lead to vertices of the class C_0. Because C_0 forms a clique, we have $\text{vis}_k(v_i) \in \text{chain}(v_{i+x_k \cdot n^\star+1}, v_{i+(x_k+1)\cdot n^\star})$. We can determine $\text{vis}_k(v_i)$, because, by Proposition 2.56, there is only one vertex of class C in $\text{chain}(v_{i+x_k \cdot n^\star+1}, v_{i+(x_k+1)\cdot n^\star})$. We now give a general method how to determine the targets of all arcs.

Let v_i be a vertex of G_{vis}. With $D_k(v_i)$ and $D_{-k}(v_i)$ we denote the set of vertices in $\text{chain}(v_i, v_{i+k})$ and $\text{chain}(v_{i-k}, v_i)$, respectively, that v_i sees. In terms of this notation, determining the arcs of G_{vis} is the same as determining $D_{\lceil n/2 \rceil}(v)$ and $D_{-\lceil n/2 \rceil}(v)$ for every vertex v. We show how to obtain $D_{\pm k}$ for all vertices by induction on k. In the same induction, we show that for any two vertices u, w of the same class, we have $|D_k(u)| = |D_k(w)|$ and $|D_{-k}(u)| = |D_{-k}(w)|$.

For $k = 1$, we can trivially determine $D_{\pm k}$ for all vertices as every vertex sees its neighbors on the boundary. By the same token, we have $|D_{\pm 1}(v)| = 1$ for all vertices v.

Now assume that, for some $k \geq 1$, we have determined $D_{\pm k}$ for all vertices, and that for any two vertices u, w of the same class, we have $|D_k(u)| = |D_k(w)|$ and $|D_{-k}(u)| = |D_{-k}(w)|$. Consider an arbitrary vertex v_i. It is sufficient to show how to infer $D_{k+1}(v_i)$ and that $|D_{k+1}(u)| = |D_{k+1}(v_i)|$ for every $u \in C_i$. Let $x := |D_k(v_i)|$. We have

$$D_{k+1}(v_i) = \begin{cases} D_k(v_i) \cup \{v_{i+k+1}\}, & \text{if } \text{vis}_{x+1}(v_i) = v_{i+k+1} \\ D_k(v_i), & \text{otherwise.} \end{cases}$$

Let $v_j := \text{vis}_{x+1}(v_i)$ be the first vertex in $\text{chain}_{v_i}(v_i, v_{i-1})$ that is not in $D_k(v_i)$. In order to derive $D_{k+1}(v_i)$, it is now enough to decide whether $v_j = v_{i+k+1}$ or $v_j \neq v_{i+k+1}$. As we know C_{i+k+1} and can infer C_j by inspecting $G^{\star}_{\text{vis}}$, this decision is trivial for $v_{i+k+1} \notin C_{v_j}$. Assume $v_{i+k+1} \in C_{v_j}$. We then have $y := |D_{-k}(v_{i+k+1})| = |D_{-k}(v_j)|$, by induction. If $v_j = v_{i+k+1}$, the arc (v_j, v_i) has index $-(y+1)$ at v_j, hence $\mathrm{LB}_{v_i}(x+1) = -(y+1)$. We want to show that $\mathrm{LB}_{v_i}(x+1) = -(y+1)$ if and only if $v_j = v_{i+k+1}$. For the sake of contradiction assume $v_j \neq v_{i+k+1}$ and $\mathrm{LB}_{v_i}(x+1) = -(y+1)$.

Let $a \in C_0$ be the first vertex from class C_0 in $\mathrm{chain}(v_i, v_j)$ and likewise let $b \in C_0$ be the last vertex from class C_0 in $\mathrm{chain}(v_i, v_j)$. Note that a and b are well defined as $v_{i+k+1} \in \mathrm{chain}(v_i, v_j)$, and hence there is a vertex of C_0 between v_{i+k+1} and v_j as $C_{v_{i+k+1}} = C_{v_j}$ (Proposition 2.56). Consider the case $a \neq b$ (cf. Figure 8.1 (left)). We define s to be the last vertex in $\mathrm{chain}(v_{i+1}, a)$ visible to v_i and t to be the first vertex in $\mathrm{chain}(b, v_{j-1})$ visible to v_j. Consider the ordered cycle $\mathcal{C} := (v_i) \oplus \mathrm{chain}(s, a) \oplus \mathrm{chain}(b, t) \oplus (v_j)$. This ordered cycle is well defined as C_0 forms a clique and hence a sees b. Note that v_i does not see any vertices in $\mathrm{chain}(b, v_{j-1})$, likewise v_j does not see any vertices in $\mathrm{chain}(v_{i+1}, a)$ (recall that $v_j = \mathrm{vis}_{x+1}(v_i)$ and $\mathrm{LB}_{v_i}(x+1) = -(y+1)$). In the subgraph induced by $\mathcal{C}$, v_i and v_j both have degree 2 which is a contradiction with Proposition 2.34. We may thus assume $a = b$ (cf. Figure 8.1 (right)). As there has to be a vertex of class C_0 in $\mathrm{chain}(v_{i+k+1}, v_j)$, there can be none in $\mathrm{chain}(v_i, v_{i+k+1})$ (and thus none in $\mathrm{chain}(v_{j-k-1}, v_j)$, as $v_{j-k-1} \in C_{v_i}$). By Proposition 2.56, this means that $k < p-2$ with $p = n/n^\star$, and thus $\mathrm{chain}(v_i, v_{i+k})$ and $\mathrm{chain}(v_{j-k}, v_j)$ do not overlap ($|\mathrm{chain}(v_{i+k+1}, v_j)| \geq p+1$, as $v_{i+k+1} \in C_{v_j}$). Let now s be the last vertex in $\mathrm{chain}(v_{i+1}, v_{i+k})$ visible to v_i and t be the first vertex in $\mathrm{chain}(v_{j-k}, v_{j-1})$ visible to v_j. We can then define the ordered cycle $(v_i) \oplus \mathrm{chain}(s, t) \oplus (v_j)$ which induces a subgraph where v_i and v_j both have degree 2, which is a contradiction to Lemma 2.56.

In conclusion, we have shown that $v_j = v_{i+k+1}$ if and only if $\mathrm{LB}_{v_i}(x+1) = -(y+1)$. We can determine $\mathrm{LB}_{v_i}(x+1)$ and y from x and the labeling of the arcs at the vertex $v_i^\star$ corresponding to C_i in $G^\star_{\mathrm{vis}}$. We thus have a necessary and sufficient criterion for deciding whether $v_j = v_{i+k+1}$ or $v_j \neq v_{i+k+1}$. The decision only depends on x and the labels of the arcs at $v_i^\star$ in $G^\star_{\mathrm{vis}}$. Let $u \in C_i$. By induction we have $|D_k(u)| = |D_k(v_i)| = x$, hence we make the same decision for u as for v_i. But this means $|D_{k+1}(u)| = |D_{k+1}(v_i)|$.

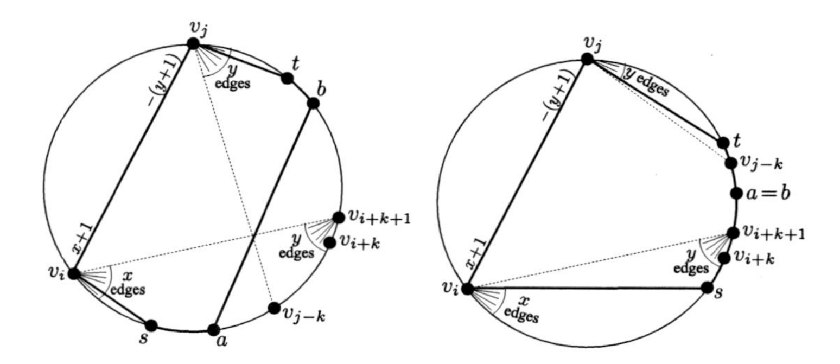

Figure 8.1.: Illustration of the visibility graph for the cases $a \neq b$ (left) and $a = b$ (right). The ordered cycle causing the contradiction is highlighted in both cases.

Algorithm 4 gives a straight forward implementation of the reconstruction algorithm that runs in time $\Theta(n^2)$, assuming $G^\star_{\text{vis}}$ is stored such that the label of an arc can be accessed in constant time (e.g., using an adjacency matrix). □

Theorem 8.6. *An agent with look-back sensor and knowledge of $\bar{n}$ can solve the visibility graph reconstruction problem.*

Proof. The theorem follows from Theorems 7.3 and 8.5. □

8.2. Reconstruction in Polynomial Time

In the previous section, we have seen that every graph in $\mathscr{F}_{\text{LB}}$ can be reconstructed from is minimum base graph in polynomial time. Using the general method for determining the minimum base graph, however, results in a very inefficient exploration strategy for the agent. In this section we develop an improved strategy for finding the minimum base graph, which achieves a polynomial running time.

The look-back sensor is quite powerful in that it allows the agent to retrace movements. In particular, the agent become capable of computing the *view* of a vertex, which is defined as follows.

Algorithm 4: Reconstruction of $G_{\text{vis}} \in \mathscr{F}_{\text{LB}}$ from $G^{\star}_{\text{vis}}$.

input : minimum base graph $G^{\star}_{\text{vis}} = (V^{\star}, A^{\star}, \lambda^{\star})$ of size $\bar{n}$
output: visibility graph $G_{\text{vis}} = (V, A)$

1 $n \leftarrow$ [maximum number of self-loops of any vertex in $G^{\star}_{\text{vis}}] \cdot \bar{n}$
2 $V \leftarrow \{v_0, v_1, \ldots, v_{n-1}\}$
3 $(C_0, C_1, \ldots, C_{n-1}) \leftarrow$ [classes along the boundary]
4 **foreach** $v_i \in V$ **do**
5     $D_1(v_i) \leftarrow v_{i+1}$
6     $D_{-1}(v_i) \leftarrow v_{i-1}$
7 **for** $k \leftarrow 2, 3, \ldots, \lceil n/2 \rceil$ **do**
8   **foreach** $v_i \in V$ **do**
9         $x \leftarrow |D_{k-1}|$
10         $C \leftarrow C_i(x+1)$
11         $y \leftarrow |D_{-(k-1)}(v_{i+k})|$
12   **if** $C = C_{i+k} \wedge \text{LB}_{v_i}(x+1) = d(C) - y$ **then**
13             $D_k(v_i) \leftarrow D_{k-1}(v_i) \cup \{v_{i+k}\}$
14             $D_{-k}(v_{i+k}) \leftarrow D_{-(k-1)}(v_{i+k}) \cup \{v_i\}$
15   **else**
16             $D_k(v_i) \leftarrow D_{k-1}(v_i)$
17             $D_{-k}(v_{i+k}) \leftarrow D_{-(k-1)}(v_{i+k})$
18 $A \leftarrow \{(u, v) \in V \times V | v \in D_{\lceil n/2 \rceil}(u)\}$
19 **return** $G_{\text{vis}} = (V, A)$

Definition 8.7. The *level-k view* of a vertex v in an arc-labeled graph G is the set of label-sequences of length at most k, for which $\Lambda(v) \neq 0$.

Lemma 8.8. *There is a returning exploration strategy for an agent with look-back sensor that, given any label-sequence Λ, determines whether $\Lambda(v) \neq 0$ in time $\Theta(|\Lambda|)$, where v is the agent's location.*

Proof. The agent can simply move according to the labels in Λ. Because of local orientation, every movement decision is unique. We have $\Lambda(v) \neq \emptyset$ if and only if Λ can be traced to its end. After finding out, the agent can retrace its movement as it is equipped with the look-back sensor. □

We can define equivalence classes with respect to views. It turns out that these classes essentially converge to the same classes we have used so far.

Definition 8.9. The *level-k class* C_v^k of a vertex v in an arc-labeled graph G is the set of all vertices with the same *level-k* view as v.

Lemma 8.10. *For every vertex v of an arc-labeled graph G and every $k \geq 1$, we have $C_v \subseteq C_v^{k+1} \subseteq C_v^k$.*

Proof. We have $C_v^{k+1} \subseteq C_v^k$ by definition. For any $u \in C_v$, we have that u and v are indistinguishable, and hence, by Proposition 2.48, $u \in C_v^{k+1}$. It follows that $C_v \subseteq C_v^{k+1}$. □

Lemma 8.11. *There is an integer n_0 for which $C_v^{n_0} = C_v$ for every vertex v of an arc-labeled graph $G = (V, A, \lambda)$.*

Proof. Consider the minimum base graph $G^\star = (V^\star, A^\star, \lambda^\star)$ of G. We let Δ be the longest label-sequence in $\{\, \delta^<_{u^\star, w^\star} \mid u^\star, w^\star \in V^\star \}$ and set $n_0 = |\Delta|$. By Lemma 8.10, we have $C_v \subseteq C_v^{n_0}$. It remains to show every vertex of $C_v^{n_0}$ is a member of C_v. For the sake of contradiction, assume there is a vertex $u \in C_v^{n_0} \backslash C_v$. Let $u^\star, v^\star \in V^\star$ denote the vertices of $G^\star$ corresponding to C_u, C_v, respectively. By assumption, $u^\star \neq v^\star$. But then, by construction, the label-sequence $\delta^<_{u^\star, v^\star}$ has length at most n_0. Because

$\delta^{<}_{u^\star,v^\star}(v) \neq \emptyset$ and $\delta^{<}_{u^\star,v^\star}(u) = \emptyset$, or vice versa, this is a contradiction to u and v having the same level-n_0 view. □

The following theorem states that the classes of a visibility graph can be computed from its level-$(n-1)$ views.

Theorem 8.12 ([39]). *Let $G_{\text{vis}} \in \mathscr{F}_{\text{LB}}$ be a visibility graph of size n. Then, for every vertex v of G_{vis}, we have $C_v^{n-1} = C_v$.*

Unfortunately, each level-k view contains a number of label-sequences that is exponential in k. In an efficient reconstruction strategy, the agent can therefore not afford to determine the level-$(n-1)$ views explicitly. The following provides an alternative way for distinguishing level-k classes by using a single label-sequence instead.

Definition 8.13. Let $G_{\text{vis}} \in \mathscr{F}_{\text{LB}}$, $k \geq 1$, and u, w be two vertices of G_{vis} with $C_u^k \neq C_w^k$. The lexicographically smallest sequence Λ of length at most k with $\Lambda(u) \neq \emptyset$ and $\Lambda(w) = \emptyset$, or vice versa, is called the *distinguishing sequence* for C_u^k and C_w^k.

distinguishing sequence

As the agent only knows an upper bound $\bar{n}$ on the number of vertices, it is inevitable to do some redundant work. To capture this, we define redundant level-k classes $\bar{C}_{v_i}^k \subseteq \{v_0, v_1, \ldots, v_{\bar{n}-1}\}$ that may contain vertices more than once. Once the agent is able to compute $G^\star_{\text{vis}}$ it can discard redundant vertices by Corollary 7.12.

Lemma 8.14. *There is a returning, polynomial exploration strategy for an agent with look-back sensor and knowledge of $\bar{n}$ that computes the redundant class $\bar{C}_v^k$, $k \leq \bar{n}$, of the agent's location v, given the set $\mathcal{D}^k$ of all distinguishing sequences for pairs of level-k classes.*

Proof. Let the agent be located at some vertex v_i. It suffices to show how the agent can decide in polynomial time whether or not $C_{v_i}^k = C_{v_{i+l}}^k$ for every $l \in [n-1]$. The level-k class is computed under the assumption that the polygon has $\bar{n}$ vertices. While this leads to redundant vertices in the output, the result is still correct when interpreting indices modulo n.

By Lemma 8.8, since $\left|\mathcal{D}^k\right| \leq \binom{\bar{n}}{2}$ and since every sequence in $\mathcal{D}^k$ has length at most $k \leq n$, the agent can determine the set $\mathcal{D}_{v_i}^+ :=$

$\{\delta \in \mathcal{D}^k \mid \delta(v_i) \neq \emptyset\}$ in polynomial time. It can then move to v_{i+l} by moving l times along the boundary in boundary order. As before, the agent can determine $\mathcal{D}^+_{v_{i+l}} := \{\delta \in \mathcal{D}^k \mid \delta(v_{i+l}) \neq \emptyset\}$ in polynomial time. By definition of $\mathcal{D}^k$, clearly $C^k_{v_i} = C^k_{v_{i+l}}$ if and only if $\mathcal{D}^+_{v_i} = \mathcal{D}^+_{v_{i+l}}$. After making this distinction, the agent can return to v_i by moving l steps along the boundary in opposed boundary order. □

It remains to show how to efficiently compute distinguishing sequences. We introduce a technical lemma that allows us to isolate computations that can be performed offline without needing the agent to move. Recall that for a class C and any vertex $v \in C$, we denote $C(x) := C_{\text{vis}_x(v)}$.

Lemma 8.15. *Let $G_{\text{vis}} \in \mathcal{F}_{\text{LB}}$ be a visibility graph of size $n \leq \bar{n}$, and let v_i be a vertex of G_{vis}. For all $k \geq 2$, the set $\mathcal{D}^k$ containing all distinguishing sequences for pairs of level-k classes, can be computed in polynomial time from the following data:*

1. *the set $\mathcal{D}^{k-1}$ of all distinguishing sequences for pairs of level-$(k-1)$ classes,*
2. *the sequence of sequences $\bar{\mathcal{C}}^{k-1}_{i+l} := (\bar{C}^{k-1}_{v_{i+l}}(1), \bar{C}^{k-1}_{v_{i+l}}(2), \ldots, \bar{C}^{k-1}_{v_{i+l}}(d(v_{i+l})))$, for every $l \in [\bar{n}]$,*
3. *the level-1 view of vertex v_{i+l}, for every $l \in [\bar{n}]$.*

Proof. We consider every pair $\{v_j, v_l\} \in \binom{\{v_i, v_{i+1}, \ldots, v_{i+\bar{n}-1}\}}{2}$ in turn and show that we can, in polynomial time, find the distinguishing sequence for $C^k_{v_j}$ and $C^k_{v_l}$ or conclude that $C^k_{v_j} = C^k_{v_l}$. The latter case can readily be detected, because $C^k_{v_j} = C^k_{v_l}$ if and only if $\bar{\mathcal{C}}^{k-1}_j = \bar{\mathcal{C}}^{k-1}_l$. It remains to show that when $C^k_{v_j} \neq C^k_{v_l}$, we can find the distinguishing sequence for $C^k_{v_j}$ and $C^k_{v_l}$.

If v_j and v_l have different level-1 views, we immediately obtain the distinguishing sequence of length one for $C^k_{v_j}$ and $C^k_{v_l}$ consisting only of the lexicographically smallest label that is present in exactly one of the views. Otherwise, $d(v_j) = d(v_l)$ and there exists an integer $q \in [d(v_j)]$, such that $\bar{C}^{k-1}_{v_j}(q) \neq \bar{C}^{k-1}_{v_l}(q)$, i.e., the q-th neighbor of v_j and the q-th neighbor of v_l belong to distinct level-$(k-1)$ classes. Since arc-labels encode boundary order, by inspecting the level-1 view of v_j, we can choose q such that the label λ of the q-th arc at v_j is smallest. There is a label-sequence $\delta \in \mathcal{D}^{k-1}$ with $\delta(v_j) \neq \emptyset$ and $\delta(v_l) = \emptyset$, or vice versa. We can find we lexicographically smallest such δ in time $O(\bar{n}^2)$ by inspecting $\mathcal{D}^{k-1}$ and we can conclude that $\lambda \oplus \delta$ is the distinguishing sequence we are looking for. $\square$

Theorem 8.16. *There is a polynomial exploration strategy for an agent with look-back sensor and knowledge of $\bar{n}$ that computes the minimum base graph $G^{\star}_{\text{vis}}$ in every graph $G_{\text{vis}} \in \mathscr{F}_{\text{LB}}$.*

Proof. Assume that there is a returning, polynomial exploration strategy that computes the redundant class $\bar{C}^{\bar{n}}_v$ of the agent's location v. The agent can move from vertex to vertex along the boundary and for each vertex execute this strategy once for the vertex itself and once for each of the vertices visible to it. This way, the agent obtains the sequence $\bar{\mathcal{C}}^{\bar{n}} := (\bar{C}^{\bar{n}}_{v_0}, \bar{C}^{\bar{n}}_{v_1}, \ldots, \bar{C}^{\bar{n}}_{v_{\bar{n}-1}})$, as well as the sequences $\bar{\mathcal{C}}^{\bar{n}}_i := (\bar{C}^{\bar{n}}_{v_i}(1), \bar{C}^{\bar{n}}_{v_i}(2), \ldots, \bar{C}^{\bar{n}}_{v_i}(d(v_i)))$, for all $i \in [\bar{n}]$. Let $\bar{\mathcal{C}}^{+}_i$ be such that the number q_i of occurrences of $\bar{C}^{\bar{n}}_{v_i}$ in $\bar{\mathcal{C}}^{\bar{n}}_i$ is maximum. By Theorem 7.11 and Lemma 8.2 it follows that C_{v_i} is a clique in G_{vis}. By Proposition 2.57, the agent can deduce $n = (q_i + 1) \cdot p$, where p is the smallest integer such that $\bar{C}^{\bar{n}}_{v_i} = \bar{C}^{\bar{n}}_{v_{i+p}}$. Once the agent knows n, it can compute $\mathcal{C}^{\bar{n}} := (C^{\bar{n}}_{v_0}, C^{\bar{n}}_{v_1}, \ldots, C^{\bar{n}}_{v_{n-1}})$, as well as the sequences $\mathcal{C}^{\bar{n}}_i := (C^{\bar{n}}_{v_i}(1), C^{\bar{n}}_{v_i}(2), \ldots, C^{\bar{n}}_{v_i}(d(v_i)))$, $i \in [n]$, from $\bar{\mathcal{C}}^{\bar{n}}$ and $\bar{\mathcal{C}}^{\bar{n}}_i$. By Lemma 8.10 and Theorem 8.12, we have $C^{\bar{n}}_v = C_v$. Thus, the agent can obtain $\mathcal{C} := (C_{v_0}, C_{v_1}, \ldots, C_{v_{n-1}}) = \mathcal{C}^{\bar{n}}$, as well as the sequences $\mathcal{C}_i := (C_{v_i}(1), C_{v_i}(2), \ldots, C_{v_i}(d(v_i))) = \mathcal{C}^{\bar{n}}_i$, $i \in [n]$. Together, $\mathcal{C}$ and $(\mathcal{C}_0, \mathcal{C}_1, \ldots, \mathcal{C}_{n-1})$ encode the arcs and vertices of $G^{\star}_{\text{vis}}$. The arc-labels can be observed by the agent during one final tour of the boundary.

It remains to show how to compute the redundant class $\bar{C}^{\bar{n}}_{v_i}$ of the agent's location v_i, returning to v_i afterwards. The agent can move from vertex to vertex along the boundary in boundary order and determine the level-1 label of each vertex and all its neighbors. After $(\bar{n} - 1)$ moves along the boundary, the agent can retrace its movements to return to v_i. At that point the time spent is polynomial in $\bar{n}$ and the agent has available a sequence $\mathcal{L}$, containing the level-1 views of the vertices $v_i, v_{i+1}, \ldots, v_{\bar{n}-1}$ as well as the sequence $\mathcal{L}_{i+l}$ of the level-1 views of the vertices visible to v_{i+l} for every $l \in [\bar{n}]$. It is now easily possible to infer $\bar{\mathcal{C}}^1 := (\bar{C}^1_{v_i}, \bar{C}^1_{v_{i+1}}, \dots, \bar{C}^1_{v_{i+\bar{n}-1}})$, as well as $\bar{\mathcal{C}}^1_{i+l} := (\bar{C}^1_{v_{i+l}}(1), \bar{C}^1_{v_{i+l}}(2), \dots, \bar{C}^1_{v_{i+l}}(d(v_{i+l})))$, for every $l \in [\bar{n}]$. We can obtain a set $\mathcal{D}^1, |\mathcal{D}^1| \leq \binom{\bar{n}}{2}$, of distinguishing sequences for every pair of level-1 classes simply by inspecting every combination of distinct views in $\mathcal{L}$. The above is the step $s = 1$ of an iterative strategy that computes $\bar{\mathcal{C}}^s := (\bar{C}^s_{v_i}, \bar{C}^s_{v_{i+1}}, \dots, \bar{C}^s_{v_{i+\bar{n}-1}})$, $\bar{\mathcal{C}}^s_{i+l} := (\bar{C}^s_{v_{i+l}}(1), \bar{C}^s_{v_{i+l}}(2), \dots, \bar{C}^s_{v_{i+l}}(d(v_{i+l})))$, and a set $\mathcal{D}^s, |\mathcal{D}^s| \leq \binom{\bar{n}}{2}$, of distinguishing sequences for every pair of level-s classes.

Assume we have described the strategy up to step $s \geq 1$. We will now describe step $s+1$. The agent first uses Lemma 8.15 to compute $\mathcal{D}^{s+1}$ from $\mathcal{D}^s$, $\mathcal{L}$, and $\mathcal{C}^s_{i+l}$ for every $l \in [\bar{n}]$. Next, the agent moves along the boundary $\bar{n}-1$ times, computing $\bar{C}^s_{v_{i+l}}$ at vertex v_{i+l} using $\mathcal{D}^{s+1}$ and the strategy of Lemma 8.14. Also, at each vertex v_{i+l}, the agent moves to each vertex $\text{vis}_q(v_{i+l})$, computes $\bar{C}^{s+1}_{v_{i+l}}(q)$ with the strategy from Lemma 8.14, and returns to v_{i+l} afterwards. As required, the agent in the end has determined $\bar{\mathcal{C}}^{s+1}, \mathcal{D}^{s+1}$, and $\bar{\mathcal{C}}^{s+1}_{i+l}$ for every $l \in [\bar{n}]$.

In summary, the agent can complete $\bar{n}$ steps of the above strategy to compute the redundant class $\bar{C}^{\bar{n}}_{v_i}$ of the agent's location v_i and return to v_i afterwards. The resulting strategy executes in polynomial time, as it involves $O(\bar{n})$ steps each of which requires polynomial time. □

Theorem 8.17. *An agent with look-back sensor and knowledge of $\bar{n}$ can solve the visibility graph reconstruction problem in polynomial time.*

Proof. The theorem follows immediately from Theorems 8.16 and 8.5. □

Theorem 8.18. *Any number of agents with look-back sensor and knowledge of $\bar{n}$ can solve the weak rendezvous problem in polynomial time.*

Proof. By Theorems 8.16 a look back agent can determine the minimum base graph $G^\star_{\text{vis}}$ in polynomial time. Using the strategy described in the proof of Theorem 8.16 the agent also obtains the distinguishing sequences for every pair of classes. From those, the agent can compute characteristic tuples, and thus find the class $\mathcal{C}^\star$ in polynomial time. By Lemma 8.14, the agent can determine the class of its current location. As it knows $G^\star_{\text{vis}}$, the agent can thus trace a path to a vertex of $C^\star$. Once every agent has executed this polynomial time strategy, weak rendezvous is established. □

Chapter 9.

Angle-Type Sensor*

In this chapter, we consider an agent with angle type sensor and knowledge of $\bar{n}$. With $\mathscr{F}_{\text{AT}} \subseteq \mathscr{F}_{\bar{n}}$ we denote a family that encodes this agent model. We develop an exploration strategy that makes use of the general tools introduced in Chapter 7. By Theorem 7.3, the agent can always infer the minimum base graph. We want to apply Theorem 7.11. For this, we need the following lemmas.

Lemma 9.1. *Let $G_{\text{vis}} = (V, A, \lambda) \in \mathscr{F}_{\text{AT}}$ have more than two classes and let $v_x, v_y \in V$ such that $C_{v_x}(2) = C_{v_y}$ and $C_{v_y}(-2) = C_{v_x}$. Then, $C_{v_{x+2}} = C_{v_y}$ and every vertex in $C_{v_{x+1}}$ is an ear.*

Proof. We first prove that for all $v_i \in V$ and $u = vis_{v_i}(2)$, we have that if $v_i = vis_u(-2)$, then $u = vis_{v_i}(2) = v_{i+2}$ and thus v_{i+1} is an ear. For the sake of contradiction assume for some $v_i \in V$ and $u = vis_{v_i}(2)$ we have $\text{vis}_u(-2) = v_i$ but $\text{vis}_{v_i}(2) \neq v_{i+2}$. Consider the ordered cycle formed by $\text{chain}(v_i, \text{vis}_{v_i}(2))$. This cycle has size at least four, as $\text{vis}_{v_i}(2) \notin \{v_{i+1}, v_{i+2}\}$. Since v_i and $\text{vis}_{v_i}(2)$ both have degree two in the subgraph induced by this ordered cycle, we have a contradiction with Proposition 2.34. Therefore, $\text{vis}_{v_i}(2) = v_{i+2}$ and v_{i+1} is an ear, as its neighbors along the boundary see each other.

Because of the above observation, it is sufficient to show that for every $v \in C_{v_x}$ we have $\text{vis}_u(-2) = v$, where $u := \text{vis}_v(2)$. For the sake of contradiction assume in the following that there is a vertex $s_0 \in C_{v_x}$ with $\text{vis}_{t_1}(-2) \neq s_0$ and $t_1 := \text{vis}_{s_0}(2)$.

We define an infinite sequence $Z = (s_0, t_1, s_1, t_2, \ldots)$ where $t_l := \text{vis}_{s_{l-1}}(2)$ and $s_l := \text{vis}_{t_l}(-2)$ for all $l > 0$. Obviously $s_l \in$

*The results presented in this chapter appeared in [11].

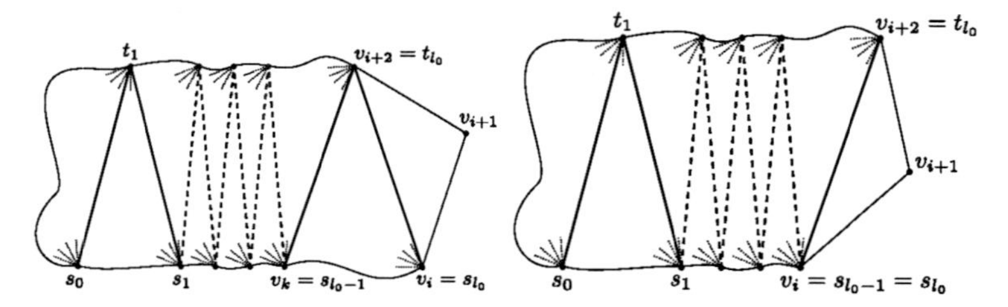

Figure 9.1.: Visualization of the "zig-zag" sequence Z. As Z does not self-intersect, there is a point l_0 from which on Z's entries do not change anymore. There are two cases how this point is reached: either s_{l_0-1} is distinct from s_{l_0} (left) or both are the same (right).

$C_{v_x}, t_l \in C_{v_y}$, for all $l \geq 0$. Intuitively, Z is the zig-zag line obtained by alternately traveling along the first and the last non-boundary arc in boundary order, starting at s_0. It is immediate to see that for any fixed index $l' \geq 0$ we have $s_l, t_l \in \text{chain}(s_{l'}, t_{l'})$ for all $l \geq l'$. Hence the part of the boundary in which these vertices lie becomes smaller and smaller, and from some index $l_0 \geq 0$ on we have $s_l = s_{l_0}$ and $t_l = t_{l_0}$, for all $l \geq l_0$. We set l_0 to be the smallest such index. Let $0 \leq i, j < n$ be such that $v_i = s_{l_0}, v_j = t_{l_0}$. We then have $\text{vis}_{v_i}(2) = v_j$ and $\text{vis}_{v_j}(-2) = v_i$. Thus, by the above observation, v_{i+1} is an ear and $v_j = v_{i+2}$. As $v_i \in C_{v_x}$ and $v_j \in C_{v_y}$, this implies $C_{v_{x+2}} = C_{v_y}$. It remains to show that every vertex in $C_{v_{x+1}}$ is an ear.

We have to consider two cases. Either s_{l_0-1} is distinct from s_{l_0} or it is the same vertex (cf. Figure 9.1). Let us assume $s_{l_0-1} \neq s_{l_0}$ and omit the discussion of the second case which is analogous. Let $0 \leq k < n$, such that $v_k = s_{l_0-1}$. As $\text{vis}_{v_k}(2) = v_{i+2}$, we have that v_k does not see any vertex in $\text{chain}(v_{k+2}, v_{i+1})$ (note that this chain is not empty as $v_k \neq v_i$). Thus, as $v_{k+1} \in C_{v_{x+1}}$ is in the same class as (the ear) v_{i+1}, the arc-labels of the arcs at v_{k+1} and at v_{i+1} encode the same angle types. Hence, the interior angle of the polygon at v_{k+1} is strictly smaller than π. For geometrical reasons (cf. Figure 9.2) no vertex in $\text{chain}(v_{i+3}, v_k)$ can see any vertex in $\text{chain}(v_{k+2}, v_{i+1})$. Let $X \subset C_{v_x}$ be the set of vertices of C_{v_x} in $\text{chain}(v_{i+3}, v_k)$ and let $Y \subset C_{v_y}$ be the set of vertices

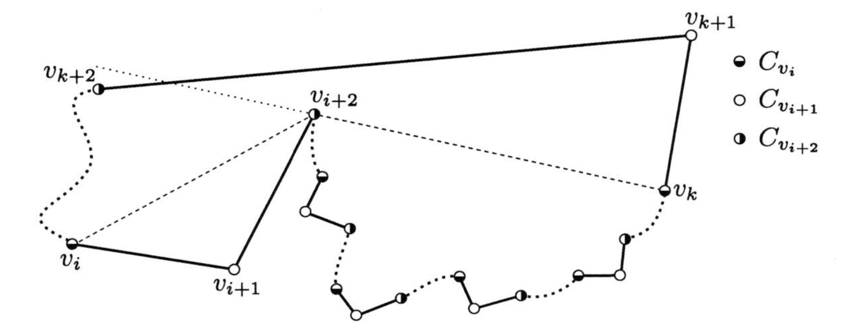

Figure 9.2.: No vertex in chain(v_{i+3}, v_k) can see any vertex in chain(v_{k+2}, v_{i+1}).

of C_{v_y} in chain(v_{i+3}, v_k). As G_{vis} has more than two classes, by Proposition 2.56, we have that $C_{v_x}, C_{v_{x+1}}, C_{v_{x+2}}$ are all different and thus X and Y are disjoint. Note that Proposition 2.56 also implies $|X| = |Y| + 1$.

We define the (undirected) bipartite graph $B_{xy} = (C_{v_x} \cup C_{v_y}, E_{xy})$ with the edge-set $E_{xy} = \{\{u, v\} \in C_{v_x} \times C_{v_y} \,|\, (u, v) \in E\}$. In B_{xy} all vertices need to have the same degree d as $|C_{v_x}| = |C_{v_y}|$ and all vertices in either class have the same degree. We have $|X| = |Y| + 1$, we have that vertices in X can only have edges to vertices in $Y \cup \{v_{i+2}\}$ and that vertices in Y can only have edges to vertices in X. For all vertices to have the same degree, v_{i+2} cannot have any edges leading to $C_{v_x} \setminus X$. This is a contradiction to the fact that v_{i+2} sees v_i which is not in chain(v_{i+3}, v_k) and thus not in X. □

Lemma 9.2. *Every graph $G_{\text{vis}} \in \mathscr{F}_{\text{AT}}$ has optimal substructure.*

Proof. Let $\mathscr{F}'_{\text{AT}} \subset \mathscr{F}_{\bar{n}}$ be the complete family that encodes the angle-type sensor and knowledge of $\bar{n}$ as described in Section 2.4.1. By definition, it is sufficient to show that every graph $G_{\text{vis}} \in \mathscr{F}_{\text{AT}}$ encodes ears and has local substructure with respect to $\mathscr{F}'_{\text{AT}}$. Let $G'_{\text{vis}} \in \mathscr{F}'_{\text{AT}}$ be a structural subgraph of G_{vis}. The arc-labels of the arcs at a vertex v of G'_{vis} can easily be obtained (cf. Section 2.4.1) from the arc-labels of the arcs at v in G_{vis}, because they encode the angle types of the angle between every pair of arcs. It remains to show that G_{vis} encodes ears, i.e., that for every ear v_i of G_{vis} we have that all vertices in C_{v_i} are ears in G_{vis}.

First, consider the case that G_{vis} has more than two classes. Since v_{i-1}, v_{i+1} see each other, we have $\text{vis}_{v_{i-1}}(2) = v_{i+1}$ and $\text{vis}_{i+1}(-2) = v_{i-1}$, and thus $C_{v_{i-1}}(2) = C_{v_{i+1}}$ and $C_{v_{i+1}}(-2) = C_{v_{i-1}}$. By Lemma 9.1, all vertices in C_{v_i} are ears. Now consider the case that G_{vis} has exactly one class. In that case, since v_i is a convex vertex, so are all vertices in C_{v_i}, as convexity is encoded in the arc-labeling. This means that the underlying polygon is convex. Thus all vertices are ears.

It remains to consider the case where G_{vis} has exactly two classes. Let $C_{v_j} \neq C_{v_i}$ be the second class in G_{vis}. Again, v_i is convex and thus all vertices in C_{v_i} are. For the sake of contradiction assume that there is a vertex $v_x \in C_{v_i}$ which is not an ear. Then v_{x-1} and v_{x+1} do not see each other, and by Proposition 2.56, $v_{x-1}, v_{x+1} \in C_{v_j}$. Let p be the Euclidean shortest path in the polygon $\mathcal{P}$ underlying G_{vis} between v_{x-1} and v_{x+1}. By Theorem 2.24, all vertices on p are reflex. This means that all vertices on p must be from C_{v_j} and thus all vertices of C_{v_j} must be reflex. Moreover, every vertex u in C_{v_j} has two neighbors u', u'' in C_{v_j} such that the angle between (u, u') and (u, u'') is reflex. If we cut off v_i from $\mathcal{P}$, we do not affect this property (every vertex u in C_{v_j} still has two neighbors from C_{v_j} forming a reflex angle) and we thus obtain a new polygon in which all vertices in C_{v_j} are still reflex, even though they might not form a class anymore. We can continue to obtain smaller and smaller subpolygons by selecting ears and cutting them off, maintaining the property that all vertices in C_{v_j} are reflex. Thus, in this process, we never cut off a vertex of C_{v_j}. This is a contradiction, as every polygon has at least one ear and thus the above process has to yield a triangle at some point. In a triangle there cannot be reflex vertices however. □

Theorem 9.3. *Any number of agents with angle-type sensor and knowledge of $\bar{n}$ can solve the weak rendezvous problem.*

Proof. The theorem follows from Theorem 7.21 together with Lemma 9.2. □

Lemma 9.4. *Let $G^\star_{\text{vis}}$ be the minimum base graph of $G_{\text{vis}} \in \mathscr{F}_{\text{AT}}$. Then, given $G^\star_{\text{vis}}$, the vertices of $G^\star_{\text{vis}}$ that represent classes of ears of G_{vis} can be identified.*

Proof. If $G^\star_{\text{vis}}$ has one vertex, the problem is trivial since G_{vis} has at least one class of ears. If $G^\star_{\text{vis}}$ has two vertices, exactly one of the two represents a class of convex vertices. This is because otherwise the underlying polygon would be convex and G_{vis} would be a clique with only one class. Since the arc-labeling of $G^\star_{\text{vis}}$ encodes convexity, it is easy to identify the vertex that represents a class of convex vertices and hence the only class of ears. Finally, if $G^\star_{\text{vis}}$ has more than two vertices, we can identify vertices that represent classes of ears using Lemma 9.1. □

Lemma 9.5. *Let $G^\star_{\text{vis}} = (V^\star, A^\star, \lambda^\star)$ be the minimum base graph of $G_{\text{vis}} \in \mathscr{F}_{\text{AT}}$. Then, given $G^\star_{\text{vis}}$, it is possible to infer the lexicographically smallest sequence $(C^{(1)}, C^{(2)}, \ldots)$ with respect to '$<$', such that $C^{(i)}$ is a class of ears in the subgraph of G_{vis} induced by $\bigcup_{j \geq i} C^{(j)}$. We write $G^{(i)}_{\text{vis}}$ to denote the subgraph of G_{vis} induced by $\bigcup_{j \geq i} C^{(j)}$.*

Proof. First of all, by Lemma 7.12, from $G^\star_{\text{vis}}$ it is possible to infer the size of G_{vis} and hence its classes. For every two vertices $u^\star, w^\star \in V^\star$, the sequence $\delta^{<}_{u^\star, w^\star}$ can be inferred by inspecting $G^\star_{\text{vis}}$ and trying out label-sequences in order of increasing lengths. Using these sequences, it is possible to evaluate the relation '$<$' for classes. By Lemma 9.4, it is possible to find the smallest class of ears $C^{(1)}$, with respect to '$<$'. Let G'_{vis} be the induced subgraph of G_{vis} obtained by cutting off the ears in $C^{(0)}$. By Lemma 9.2, we can find a minimum base graph $G^{\star\prime}_{\text{vis}}$ of G'_{vis} that encodes angle-type information by inspecting $G^\star_{\text{vis}}$. By Lemma 9.4, we can find $C^{(2)}$ similar to before and continue with the same procedure, until we remove the last vertex of a minimum base graph. □

We now introduce a counting method similar to the one used in the proof of Theorem 6.3.

Definition 9.6. Let v_i, v_h be two vertices of a polygon that do not see each other, and v_b be the first vertex other than v_i on the Euclidean shortest path from v_i to v_h. We say v_h *is hidden from* v_i *by* v_b.

Lemma 9.7. *Given $G^{\star}_{\text{vis}}$, $G^{(i+1)}_{\text{vis}} = \left(V^{(i+1)}, A^{(i+1)}, \lambda^{(i+1)}\right)$, and two vertices $v_j, v_y \in V^{(i+1)}$, it is possible to determine the number of vertices in $C^{(i)}$ hidden from v_j by v_y in the polygon underlying $G^{(i+1)}_{\text{vis}}$.*

Proof. We show the claim by induction on the number k of vertices hidden from v_j by v_y in a polygon underlying $G^{(i+1)}_{\text{vis}}$. First, observe that, by Lemma 9.5, we can compute the sequence ($C^{(1)}$, $C^{(2)}$, ...). For $k = 0$, trivially, no vertices in $C^{(i)}$ are hidden from v_j by v_y. We need to show that we can detect this situation. Observe that since $G^{(i+1)}_{\text{vis}}$ is given, we know which arc of $G^{\star}_{\text{vis}}$ corresponds to (v_y, v_j). Hence, we can count the number of arcs at v_y in $G^{(i)}$ that form a reflex angle with (v_y, v_j) by inspecting $G^{\star}_{\text{vis}}$. This number is zero if and only if $k = 0$.

Consider the case $k > 0$ (cf. Figure 9.3). Let $\mathcal{H}$ be the set of vertices in $C^{(i)}$ that are hidden from v_j by v_y. We show how to determine $|\mathcal{H}|$. Let $\mathcal{H}_{\text{vis}}$ be the set of vertices of $\mathcal{H}$ visible to v_y and $\mathcal{H}_{\text{nvis}} := \mathcal{H} \setminus \mathcal{H}_{\text{vis}}$ be the set of vertices of $\mathcal{H}$ not visible to v_y. We can easily determine $|\mathcal{H}_{\text{vis}}|$ as the number of arcs at v_y to vertices of $C^{(i)}$ that form a reflex angle with the arc (v_y, v_j). Again, we can determine this number as we know which arc of $G^{\star}_{\text{vis}}$ corresponds to (v_y, v_j) and can distinguish convex from reflex angles using the arc-labeling of $G^{\star}_{\text{vis}}$.

It remains to compute the number $|\mathcal{H}_{\text{nvis}}|$ of vertices in $C^{(i)}$ hidden from v_j by v_y and not visible to v_y. Let $\mathcal{V} \subseteq V^{(i+1)}$ denote the set of vertices that are visible to v_y and hidden from v_j by v_y. Note that since the vertices of $C^{(i)}$ are convex in the underlying polygon, they cannot hide vertices of $\mathcal{H}_{\text{nvis}}$. In fact, every vertex of $\mathcal{H}_{\text{nvis}}$ is hidden from v_y by exactly one vertex of $\mathcal{V}$. Conversely, every vertex of $C^{(i)}$ that is hidden from v_y by a vertex of $\mathcal{V}$ is part of $\mathcal{H}_{\text{nvis}}$. A vertex $v \in V^{(i+1)}$ is in $\mathcal{V}$ if and only if the arc (v_y, v) forms a reflex angle with the arc (v_y, v_j) at v_y. From $G^{(i+1)}$ and the labeling of $G^{\star}_{\text{vis}}$ we can thus identify all vertices in $\mathcal{V}$. By induction we can count the number of vertices in $C^{(i)}$ hidden from v_y by each vertex of $\mathcal{V}$ in turn. We obtain $|\mathcal{H}_{\text{nvis}}|$ as the sum of all these counts, and $|\mathcal{H}| = |\mathcal{H}_{\text{vis}}| + |\mathcal{H}_{\text{nvis}}|$. □

Lemma 9.8. *For every $i \geq 1$, it is possible to determine $G^{(i)}_{\text{vis}} = \left(V^{(i)}, A^{(i)}, \lambda^{(i)}\right)$ from $G^{\star}_{\text{vis}}$ and $G^{(i+1)}_{\text{vis}} = \left(V^{(i+1)}, A^{(i+1)}, \lambda^{(i+1)}\right)$.*

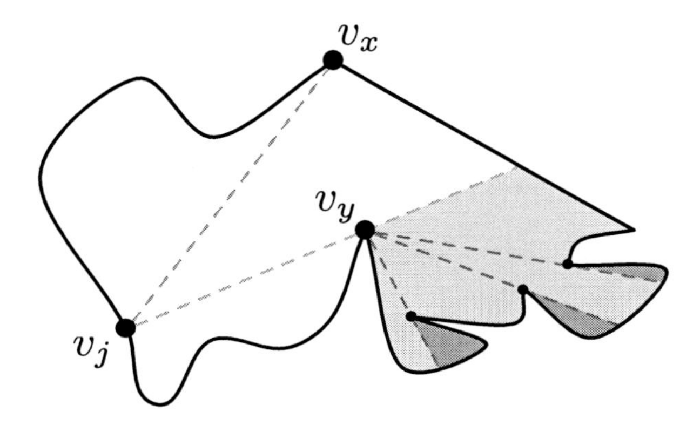

Figure 9.3.: We can count the vertices of $C^{(i)}$ hidden from v_j by v_y by counting the vertices of $C^{(i)}$ that form a reflex angle with v_j at v_y (light grey) and repeating the method recursively on the other (non-$C^{(i)}$) vertices that form reflex angles with v_j at v_y (dark grey).

Proof. Observe that, by Lemma 9.5, we can compute the sequence $(C^{(1)}, C^{(2)}, \ldots)$. The set of vertices $V^{(i)}$ of $G_{\text{vis}}^{(i)}$ is given by $V^{(i)} = C^{(i)} \cup V^{(i+1)}$. It remains to show how to construct $A^{(i)}$, as $\lambda^{(i)}$ can then be inferred from $G_{\text{vis}}^{\star}$. Let A be the set of arcs in G_{vis} between vertices of $C^{(i)}$ and $V^{(i+1)}$, and B be the set of arcs between vertices of $C^{(i)}$. We will first show how to construct A using the information available through $G_{\text{vis}}^{(i+1)}$ and $G_{\text{vis}}^{\star}$. After having determined A, we can apply the same approach in order to obtain B. This completes the proof as $A^{(i)} = A^{(i+1)} \cup A \cup B$.

Note that every arc in G_{vis} has a counterpart of opposite orientation. In order to construct A it is thus sufficient to consider $a \in V^{(i+1)} \times C^{(i)}$ and show how to decide whether $a \in A$ or $a \notin A$. Deciding which elements of $C^{(i)} \times V^{(i+1)}$ are in A is then immediate. Equivalently, we can consider $v_j \in V^{(i+1)}$ with degree d in $G_{\text{vis}}^{(i)}$ and $1 \leq k \leq d$ such that $\text{vis}_{v_j}(k) \in C^{(i)}$, and show how to "identify" $\text{vis}_{v_j}(k)$, i.e., how to find the index x such that $v_x = \text{vis}_{v_j}(k)$ in G_{vis}. If $k = 1$, we have $x = j+1$, and if $k = d$, we have $x = j-1$, because v_j sees its two neighbors on the boundary. Now assume $1 < k < d$. We will show that $v_y := \text{vis}_{v_j}(k-1)$

cannot lie in $C^{(i)}$. For the sake of contradiction, assume that $v_y \in C^{(i)}$. In any polygon underlying $G^{(i)}_{\text{vis}}$ all vertices of $C^{(i)}$ are ears and thus convex. There is more than one class in $G^{(i)}$ and thus, by Proposition 2.56, there is a vertex $v_z \in \text{chain}(v_{y+1}, v_{x-1})$ which is not visible to v_j. The Euclidean shortest path in any polygon underlying G_{vis} from v_j to v_z must visit v_x or v_y, which is a contradiction to both vertices being convex (Theorem 2.24). We can deduce that $v_y \notin C^{(i)}$ and thus $(v_j, v_y) \in A^{(i+1)}$ is part of $G^{(i+1)}_{\text{vis}}$ and has already been identified, i.e., the index y is known. Because of Proposition 2.56, it is sufficient to know how many vertices of $C^{(i)}$ are in $\text{chain}(v_{y+1}, v_{x-1})$ in order to find x. All these vertices are hidden from v_j by v_y, again because v_x is convex. Either $\text{chain}(v_{y+1}, v_{x-1})$ is empty, or all vertices hidden from v_j by v_y are in $\text{chain}(v_{y+1}, v_{x-1})$. We can distinguish these cases by inspecting $G^{\star}_{\text{vis}}$, since knowing $G^{(i+1)}_{\text{vis}}$ allows us to infer which edge in $G^{\star}_{\text{vis}}$ corresponds to (v_y, v_j). In the first case, there trivially are no vertices of $C^{(i)}$ in $\text{chain}(v_{y+1}, v_{x-1})$, and in the second case, by Lemma 9.7, we can count the number of vertices of $C^{(i)}$ in $\text{chain}(v_{y+1}, v_{x-1})$ hidden from v_j by v_y in order to determine x (cf. Figure 9.3).

Using the fact that the arcs in A have already been identified, we can apply the exact same approach to construct B. More precisely, for each $v_j \in C^{(i)}$ with degree d, and $1 < k < d$ such that $v_x := \text{vis}_{v_j}(k)$ is in $C^{(i)}$, we can infer the index x by counting the number of vertices in $C^{(i)}$ hidden from v_j by $v_y := \text{vis}_{v_j}(k-1)$. We can do this because again $v_y \notin C^{(i)}$, and because the edge $(v_j, v_y) \in A$ has been identified before. □

Theorem 9.9. *An agent with angle-type sensor and knowledge of $\bar{n}$ can solve the visibility graph reconstruction problem.*

Proof. By Theorem 7.3, the agent can determine the minimum base graph. By Lemma 9.5, the agent can compute the sequence $(C^{(1)}, C^{(2)}, \ldots)$. As the last element of $\mathcal{C}$ corresponds to a clique in the visibility graph G_{vis} and since $G_{\text{vis}} = G^{(1)}_{\text{vis}}$, the agent can determine G_{vis} using Lemma 9.8 repeatedly. □

Chapter 10.
Outlook

The main results of this thesis are summarized in Tables 10.1 and 10.2, together with some known results and open questions. We have analyzed various extensions to the basic agent model and shown that most enable the agent to solve the visibility graph reconstruction problem and often also the rendezvous problem. The only case for which we found that the agent cannot always solve the reconstruction problem is when its movements are restricted to be along the boundary only. In the case of unrestricted movements, it even remains unclear whether the basic agent model alone already empowers agents to always reconstruct the visibility graph. In fact, this is the most important question that remains open.

We motivated our work with the quest for finding *weakest* agent models that still allow the agent to map its environment. We therefore assumed a very simplistic agent model that lets agents only perceive and move to vertices. No sensing or other operations are allowed while the agent is in transit between vertices. The only additional structure that we provided was the ability to order distant vertices according to their order along the boundary. It is easy to see that without this assumption, agents cannot always solve the visibility graph reconstruction problem, even with knowledge of n (cf. Figures 4.2,4.3) and even with angle sensor (cf. Figure 5.2). It was shown prior to the results of this thesis that without any knowledge about n, the basic agent model does not empower agents to infer n, even if they have a cvv and a look-back sensor [9]. In order to show that the agent models considered in Part II of this thesis are weakest in a sense, we would need to show that an agent that only has knowledge of $\bar{n}$, but no other extensions to the basic model, cannot always solve the

	Visibility Graph Reconstruction			
	extras	result	time	Source
boundary	cvv, n	no		Theorem 4.2
boundary	cvv, n, inner-angle	no		Theorem 4.3
boundary	angle, n	yes	poly	Theorem 5.6
boundary	angle	yes	poly	Theorem 5.9
boundary	angle-type, n	*open*		
boundary	distance, n	*open*		
unrestricted	pebble	yes	poly	[50]
unrestricted	look-back, cvv	no		[9]
unrestricted	angle-type, look-back	yes	poly	Theorem 6.3
unrestricted	angle, compass	yes	poly	Theorem 6.4
unrestricted	look-back, $\bar{n}$	yes	exp	Theorem 8.6
unrestricted	look-back, $\bar{n}$	yes	poly	Theorem 8.16
unrestricted	angle-type, $\bar{n}$	yes	exp	Theorem 9.9
unrestricted	n	*open*		

Table 10.1.: Cases in which the visibility graph reconstruction problem can/cannot be solved. Running times are with respect to n or $\bar{n}$.

Weak Rendezvous			
extras	result	time	Source
optimal substructure, $\bar{n}$	yes	exp	Theorem 7.21
look-back, $\bar{n}$	yes	poly	Theorem 8.3
angle-type, $\bar{n}$	yes	exp	Theorem 9.3
n	*open*		

Table 10.2.: Cases in which the weak rendezvous problem can/cannot be solved. Running times are with respect to n or $\bar{n}$.

visibility graph reconstruction problem. Again, this remains an open problem.

In Chapter 7, we developed some general results that can help in solving this problem. By Theorem 7.3, knowledge of $\bar{n}$ is sufficient for an agent to find the minimum base graph of the visibility graph. However, we are not able to show whether or not the basic agent model with knowledge of $\bar{n}$ admits optimal substructure. Proving or disproving this might be a first step towards solving the open problem, as optimal substructure implies the existence of a class that forms a clique in the visibility graph (cf. Theorem 7.11). This, in turn, would allow agents to infer n and solve the weak rendezvous problem (Corollary 7.12, Theorem 7.21). But even then it seems difficult to infer the visibility graph from its minimum base graph. As a starting point, we give a proof that visibility graphs of convex polygons can always be reconstructed.

Lemma 10.1. *A visibility graph is regular if and only if it is complete.*

Proof. Obviously, every complete graph is regular. We claim that there is no polygon with regular visibility graph and vertices v_i, v_j, v_k, such that $v_j \in \text{chain}(v_i, v_k)$ is the only interior vertex of the Euclidean shortest path from v_i to v_k. Now, assume there is a regular visibility graph that is not complete. The visibility graph has at least one reflex vertex v_i, hence the vertices v_{i-1}, v_i, v_{i+1} contradict our claim. It remains to prove the claim.

For the sake of contradiction assume there is polygon with regular visibility graph and vertices v_i, v_j, v_k, such that $v_j \in \text{chain}(v_i, v_k)$ is the only interior vertex of the Euclidean shortest path from v_i to v_k. Choose v_i, v_j, v_k such that $\max\{|\text{LEP}(v_i, v_j)|, |\text{REP}(v_j, v_k)|\}$ is largest and among those choices take the vertices with smallest $|\text{chain}(v_i, v_k)|$. Without loss of generality, assume $|\text{LEP}(v_i, v_j)| \leq |\text{REP}(v_j, v_k)|$.

First, observe that because v_j sees v_k while v_i does not see v_k, and because v_j and v_i have the same degree, there must be another vertex v_x which is visible to v_i but not to v_j. Since v_j blocks (v_i, v_k), we have $v_x \notin \text{chain}(v_j, v_k)$. Let $p_0 = v_j, p_1, p_2, \ldots, p_m = v_x$ denote the vertices on the Euclidean shortest path from v_j to v_x. We distinguish the case $v_x \in \text{chain}(v_{k+1}, v_{i-1})$ from the case $v_x \in \text{chain}(v_{i+1}, v_{j-1})$.

Start by assuming $v_x \in \text{chain}(v_{k+1}, v_{i-1})$. Note that no vertex of $p_1, p_2, \ldots, p_m$ can lie in $\text{chain}(v_{i+1}, v_{k-1})$, since such a vertex would block (v_i, v_j) or (v_j, v_k). Similarly, no vertex of $\text{chain}(v_{x+1}, v_{i-1})$ can lie on the Euclidean shortest path, since such a vertex would block (v_x, v_j) and hence (v_x, v_i). Again, we distinguish two cases: Either v_i lies on the Euclidean shortest path, or it does not. If v_i lies on the Euclidean shortest path, it is the only vertex on the path as it sees both v_x and v_j. In this case, the triple (v_x, v_i, v_j) contradicts the choice of v_i, v_j, v_k, since $\text{REP}(v_i, v_j) \supset \text{REP}(v_j, v_k)$. If v_i does not lie on the Euclidean shortest path, we have $p_1, p_2, \ldots, p_{m-1} \in \text{chain}(v_{k+1}, v_{x-1})$. The triple (v_j, p_1, p_2) contradicts our choice of v_i, v_j, v_k, as $\text{LEP}(v_j, p_1) \supseteq \text{chain}(v_j, p_1) \supset \text{REP}(v_j, v_k)$.

Now assume we have the case $v_x \in \text{chain}(v_{i+1}, v_{j-1})$. In this case, the vertices $p_1, p_2, \ldots, p_m$ must lie in $\text{chain}(v_i, v_j)$ as they would otherwise block (v_j, v_i). If $p_1 = v_i$, the triple (v_j, v_i, v_x) contradicts our choice of v_i, v_j, v_k, since $\text{LEP}(v_j, v_i) \supseteq \text{chain}(v_j, v_i) \supset \text{REP}(v_j, v_k)$. Otherwise, the triple (p_1, v_j, v_k) contradicts our choice of v_i, v_j, v_k, since $|\text{chain}(p_1, v_k)| < |\text{chain}(v_i, v_k)|$. □

Theorem 10.2. *An agent with knowledge of $\bar{n}$ can reconstruct the visibility graph of any convex polygon.*

Proof. The visibility graph of a polygon is complete if and only if the polygon is convex. By Theorem 7.3, an agent exploring a polygon $\mathcal{P}$ can determine the minimum base graph $G^\star_{\text{vis}}$ of a visibility graph G_{vis} of $\mathcal{P}$. By inspecting the number of arcs at each vertex of $G^\star_{\text{vis}}$, the agent can determine whether G_{vis} is regular or not. By Lemma 10.1, G_{vis} is regular if and only if it is complete, or, equivalently, if $\mathcal{P}$ is convex. If $\mathcal{P}$ is convex, the agent only needs to infer n in order to reconstruct G_{vis}. This is possible as the total number of arcs in G_{vis} and hence in $G^\star_{\text{vis}}$ is exactly $n\,(n-1)$. □

The results of Chapter 4 imply that, with the basic agent model, it is not enough for agents to move along the boundary in order to reconstruct the visibility graph, even when n is known. Again, we can ask for weakest possible agent models that allow reconstruction with boundary moves only. In Chapter 5, we showed that an angle sensor is powerful enough only because of the local ordering of the angles. On the other hand, we do not know whether it would be sufficient to equip the agent with an angle-type sensor. We can also ask whether there are other agent models for which boundary movement already provides the information that the agent needs to draw a map. A very natural family of sensors that we were not yet able to obtain results for, are sensors that measure distances, for example between visible vertices. At first glance, such sensors seem quite powerful, but we were not able to uncover structural properties comparable to those used for angle sensors in Chapter 5.

Bibliography

[1] J. Abello, H. Lin, and S. Pisupati. On visibility graphs of simple polygons. *Congressus Numerantium*, 90:119–128, 1992.

[2] N. Agmon and D. Peleg. Fault-tolerant gathering algorithms for autonomous mobile robots. *SIAM Journal on Computing*, 36(1):56–82, 2007.

[3] H. Ando, Y. Oasa, I. Suzuki, and M. Yamashita. Distributed memoryless point convergence algorithm for mobile robots with limited visibility. *IEEE Transactions on Robotics and Automation*, 15(5):818–828, 1999.

[4] B. Awerbuch, M. Betke, and M. Singh. Piecemeal graph learning by a mobile robot. *Information and Computation*, 152:155–172, 1999.

[5] M. A. Bender, A. Fernández, D. Ron, A. Sahai, and S. P. Vadhan. The power of a pebble: Exploring and mapping directed graphs. *Information and Computation*, 176(1):1–21, 2002.

[6] T. Biedl, S. Durocher, and J. Snoeyink. Reconstructing polygons from scanner data. In *Proceedings of the 20th International Symposium on Algorithms and Computation*, pages 862–871, 2009.

[7] D. Bilò, Y. Disser, M. Mihalák, S. Suri, E. Vicari, and P. Widmayer. Reconstructing visibility graphs with simple robots. In *Proceedings of the 16th International Colloquium on Structural Information and Communication Complexity*, pages 87–99, 2009.

[8] P. Boldi and S. Vigna. Fibrations of graphs. *Discrete Mathematics*, 243(1–3):21–66, 2002.

[9] J. Brunner, M. Mihalák, S. Suri, E. Vicari, and P. Widmayer. Simple robots in polygonal environments: A hierarchy. In *Proceedings of the Fourth International Workshop on Algorithmic Aspects of Wireless Sensor Networks*, pages 111–124, 2008.

[10] J. Chalopin, S. Das, Y. Disser, M. Mihalák, and P. Widmayer. How simple robots benefit from looking back. In *Proceedings of the 7th International Conference on Algorithms and Complexity*, pages 229–239, 2010.

[11] J. Chalopin, S. Das, Y. Disser, M. Mihalák, and P. Widmayer. Telling convex from reflex allows to map a polygon. In *Proceedings of the 28th International Symposium on Theoretical Aspects of Computer Science*, pages 153–164, 2011.

[12] J. Chalopin, S. Das, Y. Disser, M. Mihalák, and P. Widmayer. Mapping simple polygons: How robots benefit from looking back. *Algorithmica*, to appear.

[13] V. Chvátal. A combinatorial theorem in plane geometry. *Journal of Combinatorial Theory (Series B)*, 18(1):39–41, 1975.

[14] R. Cohen and D. Peleg. Convergence of autonomous mobile robots with inaccurate sensors and movements. *SIAM Journal on Computing*, 38(1):276–302, 2008.

[15] C. Coullard and A. Lubiw. Distance visibility graphs. In *Proceedings of the 7th Annual Symposium on Computational Geometry*, pages 289–296, 1991.

[16] S Das, P. Flocchini, S. Kutten, A. Nayak, and N. Santoro. Map construction of unknown graphs by multiple agents. *Theoretical Computer Science*, 385(1–3):34–48, 2007.

[17] D. Dereniowski and A. Pelc. Drawing maps with advice. In *Proceedings of the 24th International Symposium on Distributed Computing*, pages 328–342, 2010.

[18] Y. Disser, D. Bilò, M. Mihalák, S. Suri, E. Vicari, and P. Widmayer. On the limitations of combinatorial visibilities. In *Proceedings of the 25th European Workshop on Computational Geometry*, pages 207–210, 2009.

[19] Y. Disser, M. Mihalák, and P. Widmayer. Reconstructing a simple polygon from its angles. In *Proceedings of the 12th Scandinavian Symposium and Workshops on Algorithm Theory*, pages 13–24, 2010.

[20] Y. Disser, M. Mihalák, and P. Widmayer. Reconstruction of a polygon from angles without prior knowledge of the size. Technical report, ETH Zürich, Institute of Theoretical Computer Science, 11 2010.

[21] Y. Disser, M. Mihalák, and P. Widmayer. A polygon is determined by its angles. *Computational Geometry: Theory and Applications*, 44:418–426, 2011.

[22] B. R. Donald. On information invariants in robotics. *Artificial Intelligence*, 72(1-2):217–304, 1995.

[23] H. Everett. *Visibility graph recognition.* PhD thesis, University of Toronto, Department of Computer Science, January 1990.

[24] H. Everett and D. G. Corneil. Negative results on characterizing visibility graphs. *Computational Geometry: Theory and Applications*, 5:51–63, 1995.

[25] S. Fisk. A short proof of Chvatal's watchman theorem. *Journal of Combinatorial Theory (Series B)*, 24(3):374, 1978.

[26] P. Fraigniaud, D. Ilcinkas, G. Peer, A. Pelc, and D. Peleg. Graph exploration by a finite automaton. In *Proceedings of the 29th Symposium on Mathematical Foundations of Computer Science*, pages 451–462, 2004.

[27] A. Ganguli, J. Cortés, and F. Bullo. Distributed deployment of asynchronous guards in art galleries. In *Proceedings of the 2006 American Control Conference*, pages 1416–1421, 2006.

[28] B. Gfeller, M. Mihalak, S. Suri, E. Vicari, and P. Widmayer. Counting targets with mobile sensors in an unknown environment. In *Proceedings of the 3rd International Workshop on Algorithmic Aspects of Wireless Sensor Networks*, pages 32–45, 2007.

[29] S. K. Ghosh. On recognizing and characterizing visibility graphs of simple polygons. *Discrete and Computational Geometry*, 17:143–162, 1997.

[30] S. K. Ghosh. *Visibility Algorithms in the Plane*. Cambridge University Press, 2007.

[31] S. K. Ghosh and P. P. Goswami. Unsolved problems in visibility graph theory. In *Proceedings of the India-Taiwan Conference on Discrete Mathematics*, pages 44–54, 2009.

[32] L. Jackson and S. K. Wismath. Orthogonal polygon reconstruction from stabbing information. *Computational Geometry*, 23(1):69–83, 2002.

[33] C. Jordan. *Cours D'Analyse*. Gauthier-Villars, 1893.

[34] M. Katsev, A. Yershova, B. Tovar, R. Ghrist, and S. M. LaValle. Mapping and pursuit-evasion strategies for a simple wall-following robot. *IEEE Transactions on Robotics*, 27(1):113–128, 2011.

[35] A. Komuravelli and M. Mihalák. Exploring polygonal environments by simple robots with faulty combinatorial vision. In *Proceedings of the 11th International Symposium on Stabilization, Safety, and Security of Distributed Systems*, pages 458–471, 2009.

[36] D.-T. Lee and F. P. Preparata. Euclidean shortest paths in the presence of rectilinear barriers. *Networks*, 14(3):393–410, 1984.

[37] J. Lin, A. Morse, and B. Anderson. The multi-agent rendezvous problem. part 1: The synchronous case. *SIAM Journal on Control and Optimization*, 46(6):2096–2119, 2007.

[38] J. Lin, A. Morse, and B. Anderson. The multi-agent rendezvous problem. part 2: The asynchronous case. *SIAM Journal on Control and Optimization*, 46(6):2120–2147, 2007.

[39] N. Norris. Universal covers of graphs: isomorphism to depth $n-1$ implies isomorphism to all depths. *Discrete Applied Mathematics*, 56(1):61–74, 1995.

[40] J. M. O'Kane and S. M. LaValle. Dominance and equivalence for sensor-based agents. In *Proceedings of the 22nd National Conference on Artificial Intelligence and the 19th Innovatice Applications of Artificial Intelligence Conference*, pages 1655–1658, 2007.

[41] J. O'Rourke. Uniqueness of orthogonal connect-the-dots. In G. T. Toussaint, editor, *Computational Morphology*, pages 97–104. North-Holland, 1988.

[42] J. O'Rourke and I. Streinu. The vertex-edge visibility graph of a polygon. *Computational Geometry*, 10(2):105–120, 1998.

[43] P. Panaite and A. Pelc. Exploring unknown undirected graphs. *Journal of Algorithms*, 33:281–295, 1999.

[44] D. Rappaport. On the complexity of computing orthogonal polygons from a set of points. Technical Report SOCS-86.9, McGill University, Montreal, Canada, 1986.

[45] O. Reingold. Undirected st-connectivity in log-space. In *Proceedings of the 37th Annual ACM Symposium on Theory of Computing*, pages 376–385, 2005.

[46] A. Sidlesky, G. Barequet, and C. Gotsman. Polygon reconstruction from line cross-sections. In *Proceedings of the 18th Annual Canadian Conference on Computational Geometry*, pages 81–84, 2006.

[47] J. Snoeyink. Cross-ratios and angles determine a polygon. *Discrete and Computational Geometry*, 22(4):619–631, 1999.

[48] G. Srinivasaraghavan and A. Mukhopadhyay. A new necessary condition for the vertex visibility graphs of simple polygons. *Discrete and Computational Geometry*, 12(65–82), 1994.

[49] I. Streinu. Non-stretchable pseudo-visibility graphs. *Computational Geometry: Theory and Applications*, 31(195–206), 2005.

[50] S. Suri, E. Vicari, and P. Widmayer. Simple robots with minimal sensing: From local visibility to global geometry. *International Journal of Robotics Research*, 27(9):1055–1067, 2008.

[51] I. Suzuki and M. Yamashita. Distributed anonymous mobile robots: Formation of geometric patterns. *SIAM Journal on Computing*, 28(4):1347–1363, 1999.

[52] O. Veblen. Theory of plane curves in non-metrical analysis situs. *Transactions of the American Mathematical Society*, 6(1):83–98, 1905.

[53] M. Yamashita and T. Kameda. Computing on anonymous networks: Part I – characterizing the solvable cases. *IEEE Transactions on Parallel and Distributed Systems*, 7(1):69–89, 1996.

Curriculum Vitae

Yann Disser

born on March 4, 1983, in Frankfurt a.M., Germany

2008 – 2011	**PhD in Theoretical Computer Science** ETH Zürich
2006 – 2008	**MSc in Physics** TU Darmstadt (*grade: with distinction*)
2003 – 2007	**Diploma in Computer Science** TU Darmstadt (*grade: with distinction*)
2005 – 2006	**Study Abroad** University of Saskatchewan (*Dean's Honour List*)
2003 – 2006	**BSc in Physics** TU Darmstadt (*grade: very good*)
2002 – 2003	**Community Service** Gustav-Heinemann-Schule für Praktisch Bildbare, Dieburg
1999 – 2002	**Abitur** Alfred-Delp-Schule Dieburg (*grade: 1.1*)